Mathematics
for the
Curious

Mathematics
for the
Curious

Peter M. Higgins

Oxford   New York

OXFORD UNIVERSITY PRESS

1998

Oxford University Press, Great Clarendon Street, Oxford OX2 6DP

Oxford New York

Athens Auckland Bangkok Bogota Bombay
Buenos Aires Calcutta Cape Town Dar es Salaam
Delhi Florence Hong Kong Istanbul Karachi
Kuala Lumpur Madras Madrid Melbourne
Mexico City Nairobi Paris Singapore
Taipei Tokyo Toronto Warsaw

and associated companies in
Berlin Ibadan

Oxford is a trade mark of Oxford University Press

British Library Cataloguing in Publication Data
Data available

Library of Congress Cataloging in Publication Data
Higgins, Peter M., 1956–
Mathematics for the curious / Peter M. Higgins.
Includes index.
1. Mathematics — Popular works. I. Title.
QA93.H47 1998 510—dc21 97–40546 CIP
ISBN 019-288072-1 (pbk.)

1 3 5 7 9 10 8 6 4 2

Typeset by Jayvee, Trivandrum, India
Printed in Great Britain by
Cox & Wyman,
Reading, England

Preface

This book is meant for enjoyment and so you, the reader, should have no inhibitions dipping into it wherever you fancy. There will be occasional references to things that appear earlier, but there will be no great loss if you ignore such comments and read on. However, you may find it just as satisfying to read the book by travelling back and forth through it. This may sound disorganized, but it is the way people on the whole learn about maths.

I would like to thank those who have helped read drafts of the book, the staff and anonymous readers of Oxford University Press, and also Genevieve Higgins and Dr Tim Lavers for their proofreading and valuable comments.

P.M.H.

Colchester, July 1997

Contents

1

Ten Questions and their Answers

Many things in the world have a mathematical side to them and it is the business of mathematics to try to understand that facet of their nature. Mathematical reasoning can often explain things that otherwise remain obscure or baffling, and sometimes the reasoning involved is easy to understand once it is presented to you.

This introductory chapter consists of a series of examples meant to prove my point. If you feel yourself any the wiser after reading them through, I invite you to read on. This book makes no claim to great depth, but through it I hope to convey the flavour of modern mathematics. It may also serve to clarify some aspects of school algebra and geometry, and even arithmetic, about which you may have always felt somewhat uneasy. It is perfectly possible, for instance, for anyone to understand the theorem of Pythagoras as well as a professional mathematician; the level of difficulty encountered is only that met when assembling a jigsaw puzzle of half a dozen pieces. There is no reason why such genuinely interesting aspects of mathematics need remain a mystery—indeed, most thoughtful people, with a little patience, can understand such things completely. Even some profound aspects of twentieth-century mathematics are quite accessible, and I hope to give the reader the satisfaction of seeing some parts of the mathematical world that were never revealed to even the greatest minds of the past.

At school, and equally at university, the student and teacher are working mainly towards satisfactory examination marks and there is often no time to wonder at the mathematical scenery.

This is not the case with us, however. The reader has no one to please but him- or herself. There is no hurry nor threat of judgement. Take time to ponder over what arises. A pencil and paper might help on occasion, and you need have no inhibition as to what you scribble and draw. Although such doodlings can look childish and useless, they are a real aid to the thought process and are never to be despised.

1. How many matches are played in a tennis tournament?

This is a practical question to which the tournament organizers would certainly need to know the answer. Let us take, for example, a typical Grand Slam tournament where there are 128 entrants. Each round consists of pairing off the remaining players. Each player then plays the opponent with whom he or she has been drawn. The losers retire from the tournament while the winners progress to the next round until the champion is decided.

This is not a difficult problem to solve. Clearly, there are $128 \div 2 = 64$ matches in the first round, leaving 64 players to fight out the second. The next round will then require $64 \div 2 = 32$ matches, and so on. The total number of matches in the tournament is then seen to be:

$$64 + 32 + 16 + 8 + 4 + 2 + 1 = 127.$$

The problem is solved, but there seems to be the seed of something interesting in the answer itself—127, exactly one less than the total number of players. What is going on?

First we might notice that the number of entrants is a rather curious figure itself—128. The organizers have shrewdly chosen a number that is a power of 2: $128 = 2^7$ meaning $2 \times 2 \times 2 \times 2 \times 2 \times 2 \times 2$. The effect of this is to ensure that at the completion of any round there is an even number of players remaining who can then be conveniently paired off (except of course after the final round, where just one remains undefeated). Someone blessed with knowledge of so-called geometric series might now make the point, quite correctly, that we have just checked that $2^6 + 2^5 + 2^4 + 2^3 + 2^2 + 2^1 + 1 = 2^7 - 1$, which is merely a special case of the formula concerning sums of powers of 2, namely that, for any whole number n,

$$2^n + 2^{n-1} + 2^{n-2} + \ldots + 2^2 + 2^1 + 1 = 2^{n+1} - 1.$$

I hasten to assure you that this somewhat technical observation actually misses the point. What point? To see what I am getting at, let us alter the example slightly. Suppose we naively allow 100 entrants rather than 128 into our tournament. This might happen in an amateur setting where we simply wish to grant everyone who wants to play the right to enter. Clearly, there are 50 matches in the first round and 25 in the second; but then we are left with an *odd* number of players, 25. To deal with this, we rule that in such circumstances we choose one of the remaining players at random who is given a bye—that is a free passage to the next round. The other players are then paired off to play in the usual way. Under these rules there will be 13 players remaining after 3 rounds (12 winners of the third-round matches together with the player who enjoyed the bye). A little thought reveals that the total number of players remaining at the beginning of each round is given by the sequence 100, 50, 25, 13, 7, 4, 2, and the total number of matches in our tournament is given by the sum:

$$50 + 25 + 12 + 6 + 3 + 2 + 1 = 99.$$

Again we have the answer, although arriving at the final figure was a little awkward. If you repeat the calculation for different numbers of entrants you will find that, beginning with n players, you always have to schedule $n - 1$ matches. There must be a reason for this—a proof if you like. An argument based on dividing by 2 and adding a certain number of times seems fraught with difficulty. One is liable to bump into odd numbers of remaining players at quite irregular intervals, as we did here, and it would seem very difficult to describe the entire process in general with sufficient accuracy to reach the required conclusion, tantalizingly simple though it is.

This kind of thing often happens to mathematicians. They feel sure of a certain proposition, and they have what at first sight seems a natural approach to establishing it, but they then run into difficulties in completing their proof. The line of attack to the problem can let you in for more than you bargained for and you are apparently forced to deal with aspects of the situation with which you were not initially interested at all. Frequently, as in this case, we have failed to see the wood for the trees. What is required is for us to step back in order to spot what is going on. The key observation is that *the number of matches is exactly the same as the number of losers*: each match produces one loser and every player, except the eventual champion, will lose just the once. Therefore there must always be one less match than there are players.

This is a beautifully clean proof—it goes right to the heart of the matter and allows us to understand what is going on by furnishing a clear-cut reason why the observed result is always as it is. Although simple and short, this kind of reasoning is by no means easy to devise, so there is no shame in not seeing it oneself. If you did happen to see it coming, you have cause to congratulate yourself.

This principle of identifying a one-to-one correspondence between some set of interest and another set which is relatively simple to count (see Figure 1) arises constantly in the theory of enumeration and probability. Simple as it looks, this really is a

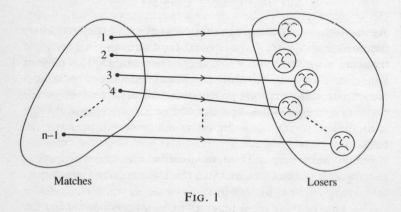

Matches   Losers

Fig. 1

big idea. One has the right to take the view, at least if you saw the solution very quickly, that I am overstating my case, as the idea is terribly obvious. However, I know that intelligent people can often stare at the above problem for hours without ever quite spotting the crucial correspondence. Mathematicians might expect to have less trouble, but mainly because they have learnt the tricks of the trade. It really is not obvious, merely simple, and so it is easy enough once you are shown.

A similar problem concerns sharing a rectangular chocolate bar.

2. What is the minimum number of snaps required to break a chocolate bar into its individual pieces?

Let us suppose that we have a 4×5 chocolate block with 20 pieces. The answer to the question is 19. Why? Because each time you snap a piece of the chocolate the total number of pieces increases by one. Since you began with one piece you will need to snap the bar 19 times to create the 20 individual pieces.

There is often much to be learnt from a problem after you have solved it. Remember, you are not trying to build up points on some examination. You are trying to learn something about mathematics. Much may be gleaned by taking a moment to reflect on what you have just seen.

First, the fact that the bar was rectangular did not enter into the solution. It could have been a single block of any shape. Second, we solved the problem for a 20-square block, but clearly the argument is valid for any number of squares; in general, if there are n squares, then $n - 1$ snaps are required. This is what mathematics is about, finding general results and principles rather than solving a problem for special values such as 20 in this case. You will come across this theme of working from the particular to the general on many occasions throughout the book.

Finally, we were asked for the minimum number of snaps, but our argument actually shows it to be the maximum number as well! However one goes about it, $n - 1$ snaps will always yield n pieces. The result is independent of the way the process is carried out. There is no especially clever way of doing it. This may be disappointing, but it is well worth knowing. This is one of the uses of mathematics—it often tells you when you are wasting your time trying to do the impossible.

Our third problem, concerning a clock face, we shall solve in three ways.

3. When do the hands of the clock coincide?

Let us be specific. What is the exact time, after 12 noon, when the minute hand of a clock lies directly over the hour hand? (See Figure 2(*a*).) We can all see immediately that it is some time a little after 1.05, but when exactly?

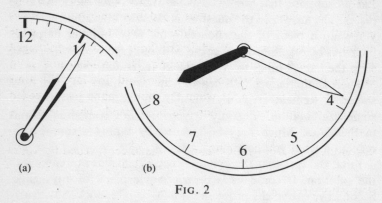

(a)   (b)

Fɪɢ. 2

Here is a quick solution. After the passing of the stroke of noon, up till and including midnight, there are 11 occasions when the hands coincide in this fashion (yes 11, not 12). These are all equally spaced in time, so the time between successive coincidences must be $12 \div 11 = 1\frac{1}{11}$ hours, which is, to the nearest second, 1 hour 5 minutes and 27 seconds.

This is a slick solution which exploits the symmetry inherent in the question: successive pairs of coincides are equally far apart. However, it can leave the uneasy feeling that mathematical wool has been pulled over our eyes. The count of the coincidences requires some thought, and the manner in which we left out one endpoint (noon) and included the other (midnight) may seem a bit dubious. It might be clearer to begin from some other point—say 1 o'clock—and check that, indeed, 11 coincides occur in the next 12-hour period. This avoids any apparent difficulty arising from the endpoints of the time interval we use.

However, it is a pretty problem, so let us solve it again. This time we look at it differently. Let us imagine that the tips of the minute hand and of the hour hand respectively represent two rather leisurely runners trotting around a circular track. The minute hand completes exactly one circuit per hour, while the hour hand crawls along at a snail's pace at $\frac{1}{12}$ of a circuit per hour. The question now is, when does the minute hand first lap the hour hand?

We translate the problem into a very simple equation as follows. After t hours the minute hand has run exactly t circuits of the clock face while the hour hand has gone only $\frac{t}{12}$ times around; for instance, if $t = 4$, then the minute hand has gone around exactly four times while the hour hand has managed only $\frac{4}{12} = \frac{1}{3}$ of a circuit—it is 4 o'clock. The problem now is to find the value of t at which the minute hand first laps the hour hand. This will be the time when the minute hand has travelled exactly one full lap more than the hour hand. This yields the equation:

distance travelled by minute hand = distance travelled by hour hand $+ 1$;

that is to say:

$$t = \frac{t}{12} + 1.$$

If we subtract $\frac{t}{12}$ from both sides of this equation we obtain $\frac{11}{12}t = 1$, or in other words we recover our original solution: $t = \frac{12}{11} = 1\frac{1}{11}$ hours.

This approach can be used to solve any problem of this kind. For instance, non-working display clocks in jewellery shops are often fixed at the time approaching 20 past 8, when the hour and minute hands are equal distances from the numeral 6 on the clock face (see Figure 2(*b*)). The exact moment when this symmetry occurs turns out to be $18\frac{6}{13}$ minutes after 8. The equation we need is a little more complex this time. For each minute that passes, the minute hand sweeps out $\frac{360}{60} = 6°$ while the hour hand passes through only one-twelfth of this angle: $\frac{6}{12} = \frac{1}{2}$ of one degree. It follows that t minutes after 8 the angle between the minute hand and a line passing from the centre of the face through the numeral 6 is given by $180 - 6t$ degrees. The corresponding angle made by the hour hand begins at $60°$ and increases by $\frac{t}{2}$ degrees each minute. We want the value of t when these two angles agree; that is to say, we require the solution to the equation:

$$180 - 6t = 60 + \frac{t}{2}.$$

This yields

$$\frac{13t}{2} = 120,$$

giving the value for t of $18\frac{6}{13}$ minutes mentioned above. To the nearest second, this is 18 minutes and 28 seconds after 8 o'clock.

Our final solution to the problem is the least slick and mathematically the most complicated. Nevertheless, it uses the approach that most people seem naturally to adopt to the problem—the Achilles and Tortoise Technique. Since one can see the approximate answer almost immediately, it is practically irresistible to proceed by successive approximations as follows. The first coincidence occurs soon after 1 o'clock, so imagine the hands standing at that time. Five minutes later, Achilles (the minute hand) has reached the numeral 1 on the clock. However, Tortoise (the hour hand) has, in the meantime, moved on a little—to be exact, since $\frac{1}{12}$ of one hour has passed and Tortoise moves at $\frac{1}{12}$ of a circuit each hour, Tortoise has moved $\left(\frac{1}{12} \times \frac{1}{12}\right)$ of an hour forwards. It will then take Achilles this amount of time to reach the point where Tortoise is now, by which time Tortoise will have moved $\left(\frac{1}{12} \times \frac{1}{12} \times \frac{1}{12}\right)$ of an hour forwards, and so on.

This does not look like a solution at all, but merely a sequence of better and better approximations. Indeed, the Greeks of antiquity thought that this approach led to insurmountable difficulties, as it produced an unending list of tasks for Achilles to perform before he could catch Tortoise. This would take an infinite amount of time, so poor Achilles could never catch Tortoise. This was one of Zeno's Paradoxes.

There is no need for confusion here. The fact that we have mentally contemplated a finite time interval as an infinite collection of tiny intervals will not cause Achilles any difficulty. The false assumption that leads to the paradox is that, as we sum an unending sequence of positive numbers, the successive sums we calculate must increase beyond all bounds. This may sound reasonable, but we have shown it to be false. Since we have already solved this problem—twice in fact—we can conclude that:

$$1 + \frac{1}{12} + \left(\frac{1}{12}\right)^2 + \left(\frac{1}{12}\right)^3 + \ldots = 1\frac{1}{11}.$$

What does this mean? It means that $1\frac{1}{11}$ is the least number that is not exceeded by adding some finite number of terms from this infinite sequence. If you like, the sums obtained from this series by adding more and more terms grow larger and larger yet never exceed the limiting value of $1\frac{1}{11}$.

This is, incidentally, another example of a geometric series. We somewhat accidentally encountered one in our first question concerning the tennis tournament. (There it involved powers of 2 and only a finite number of terms.) We shall say more about this important class of series in a later chapter, but to satisfy your curiosity on the matter temporarily I will state in passing that a series of the form:

$$1 + r + r^2 + r^3 + \ldots,$$

where r is a positive number less than 1, has the limiting value of $1/(1 - r)$. In our example $r = \frac{1}{12}$; a simpler example is given by taking $r = \frac{1}{2}$; our (unproved) summation formula then gives:

$$1 + \frac{1}{2} + \frac{1}{4} + \frac{1}{8} + \frac{1}{16} + \ldots = 1 \Big/ \left(1 - \frac{1}{2}\right) = 1 \Big/ \frac{1}{2} = 2.$$

Our clock question has proved quite fruitful. Our fourth problem is, if anything, easier.

4. Does a 10% cut followed by a 10% increase have no overall effect?

A workman, who is paid at an hourly rate, has his pay cut by 10% while the number of hours he is to work is increased by 10%. His boss assures the worker that this is for his own good, as the boss has been reluctantly forced to reduce the pay rate in order to remain competitive and has increased the required number of hours as a favour:

'You do earn 10% less per hour, but you have 10% more hours so that your weekly pay will be maintained.'

Is this true? Let us suppose that the workman earned £100 per week before. He has had his hourly rate cut by 10% so the effect of that in itself will be to reduce his weekly earnings to £90. The number of hours has increased by 10%. Now 10% of 90 is 9, so this will give him an extra £9 per week, taking his pay packet back up to £99—not £100!

Might it help if we do the calculation the other way around? Imagine that we increase his hours first by 10%: that will take his pay up to £110. Then we decrease the pay rate by 10%: 10% of 110 is 11; that takes his pay back down to £99. Whichever way we look at it, he is a loser. This looks very unfair—mathematics seems to be conspiring with the boss to exploit the workers. In any case, what is wrong with the boss's argument?

The argument of the boss is flawed, and the flaw lies in his not identifying the subject when he loosely talked of 10%. If you decrease by 10% and then increase *what you now have* by the same percentage, you don't get back to where you started; whatever order you do it in, the effect is an overall 1% decrease.

There seems to be a disturbing lack of symmetry here. Let us look at another similar problem to see if the balance can be restored. Suppose that our worker received instead a 10% increase in his hourly rate and at the same time took a 10% reduction in the number of working hours. Perhaps this is the other side of the coin—his overall pay will now go up to £101? I'm afraid this is wishful thinking. However you look at it, he winds up with £99 again, although under these rules he has more time left in which to spend it. (The calculations are just the same as before—try them yourself.)

Again, it is not too hard to see through the mystery. Let P represent the worker's weekly pay. In each case two actions are taken: one has the effect of increasing his pay by 10%, that is multiplying P by $1 + 0.1 = 1.1$; the other decreases his pay by the same percentage, which is achieved by multiplying by $1 - 0.1 = 0.9$. Now, the order in which we do these multiplications does not matter (it never does):

$$P \times 1.1 \times 0.9 = 0.99 \times P = P \times 0.9 \times 1.1,$$

and so our poor worker will always end up 1% worse off for such a pair of changes. Notice that we are generalizing again, and for good reason—examining the general situation makes the phenomenon clearer, as we are less distracted by particular aspects of the case (in this instance the actual value of P) which have little bearing on the underlying problem.

Problems and arguments involving percentages cause much confusion and cannot be ignored as they pervade all our dealings with money and many other things besides. But what is a per cent, and why do percentages arise so frequently?

The first part of the question is easily answered: 1 per cent of a quantity is simply a $\frac{1}{100}$th part. Why do we then put such emphasis on this particular fraction?

The answer is a purely pragmatic one, and for that reason not of much interest to mathematicians (although problems involving compound interest begin to have some real mathematical substance and lead to questions that are not just matters of simple arithmetic—but of this, more later).

That we work with a power of 10 $(100 = 10^2)$ is because the base of the number system we have chosen is 10 (undoubtedly because of the number of fingers we have). As you may know, computer systems often use base 2, or binary, so that only two symbols, 0 and 1, are required, corresponding to the two natural machine states of 'off' and 'on'. It has often been argued that arithmetic would be much easier, and so better understood, if we worked in a duodecimal system with 12 as our base, because 12 is a number with more factors than 10. This would force us to introduce two new symbols for the numbers 10 and 11 but base 12 would be just as usable as base 10. For instance, the number 171 in base 12 becomes 123, as the meaning of 123 in base 12 is $1 \times 12^2 + 2 \times 12 + 3$. Any number ending in a 3 in base 12 is a multiple of 3, since 3 is a factor of 12. (This exactly matches the situation where every number ending in a 5 in base 10 is a multiple of 5.) It is not so obvious that 171 is a multiple of 3 when written in base 10 (although it is easy to check, as any number is a multiple of 3 if the sum of its digits, 9 in this case, is a multiple of 3).

Like Esperanto, a duodecimal system would undoubtedly have much to recommend it if the world could cope with the change. Logical though it may be, the duodecimal system seems for ever doomed to remain a fundamentally sound idea that will never be adopted.

Why do we give a special name to a $\frac{1}{100}$th part rather than to $\frac{1}{10}$ or $\frac{1}{1000}$? Small numbers are easier to think about than large ones, so that is the reason why 1000 is rejected. The final practical advantage that 100 has over 10 is that, as a rule of thumb, a $\frac{1}{100}$th part of some quantity is about the smallest part that is significant: a 1% cut in our salary is just large enough for us to feel really annoyed. It is therefore natural to give a special name, 'per cent', to a $\frac{1}{100}$th part. The effect of this is that, in most discussions concerning quantities of money in particular, the numbers involved will be of a nice manageable size—numbers that can be counted on fingers and toes.

This practical point, the actual size of units, is a factor that is often overlooked when the benefits of one system of units over another is debated. For instance, I am sure that almost everyone assumes that people of a scientific bent who are naturally adept with numbers will automatically prefer the metric system of measurement to the imperial. The matter is not, however, so clear-cut. Both systems have relative advantages and disadvantages. The metric system has units based on powers of 10, making it consistent with base 10 arithmetic which confers obvious advantages in calculation. There is also another consistency, in that the unit of volume, the litre, is chosen so that 1 litre of pure water has the mass of 1 kilogram. This is also a practical benefit.

The size of the metre, though, is extremely arbitrary, one might even say silly, being $\frac{1}{10,000,000}$ the distance from the equator to the North Pole. This makes it a natural unit in a way, but not in a way that is of much real use.

On the other hand, the actual sizes of the imperial units of inch and foot are very practical because they are attuned to the human scale of things in a way that the centimetre (too small) and metre (a bit too large) are not. Humans are about 5–6 feet tall and have hands that are 6–8 inches long. Consequently they surround themselves with objects of similar scale, which are then conveniently measured in units of feet and inches. Another reason why the old units are liked is that the very words 'feet' and 'inches' are shorter and trip more easily from the tongue than do 'centimetre' and the like. Although an entirely linguistic point, this is none the less significant—phrases such as 'missed by a mile' and 'inching along' are useful expressions which don't translate to metric in a fashion that lends itself to normal usage. If there had been 10 inches to the foot, the imperial system might have conquered the world.

Just as it is possible for people to learn two languages, it is perfectly possible to usefully employ two systems of measurement. I hope that both systems survive for a long time and that we learn to be relaxed about this. There is no reason to declare one or other of them heretical. They are both part of our culture.

Talking of a 'part' or 'parts' makes sense only when we know what is the underlying object of discussion. As we saw in the example, confusion and ambiguity can arise when we allow ourselves to speak of 10% as if it meant something in isolation—it means 10% *of* something, and we need to know what that something is.

A common claim is that you cannot have more than 100% of anything and so any statement involving percentages larger than 100 is inherently nonsense. Certainly a statement such as 'the price of shares in Fabtex has fallen by 150%' is nonsense. However, Fabtex shares could well rise in price by 150%: this simply means that the price *increase* is equal to 1.5 times the original price.

Recently a television reporter was brought to task when he said that the rate of unemployment in a town had increased from 20% to 25%—an increase of 5%. Apparently people did ring the television station to point out that, if a quantity increases from 20 units to 25 units, the percentage increase is

$$\frac{25 - 20}{20} \times 100 = 25\%;$$

in other words, the increase is equal to one-quarter of the original number. In this case the quantity in question is itself a percentage, but no matter: that quantity has increased by 25%, not 5%. The commentator's remark does of course have meaning, in that the increase in the number of unemployed is equal to 5% of the town's workforce. Again, it is simply a matter of our agreeing as to what we are referring to when we speak of a certain percentage.

As I have conceded, our problem of the worker's wages, curious though it might seem, is from a strict mathematical viewpoint the least interesting of our questions so far, although I have seen an instance of a student quite struck by what amounts to the same thing. She noticed that if you take any number, such as 10, and multiply the two numbers either side of it together, in this case 9 and 11, you always get exactly one less than the square of the original number; in this case, $9 \times 11 = 99 = 100 - 1$.

'Isn't that amazing?' asked she rhetorically. Much as I was reluctant to do anything to dampen enthusiasm in a student, I had to tell her that it wasn't as stunning as she imagined, and could be instantly explained. All that she had noticed was that, for any number n:

$$(n - 1)(n + 1) = n^2 - 1.$$

Anyone who still remembers their secondary school algebra will be able to multiply out the brackets on the left and verify this little identity. Again, we see that something that may seem mysterious and difficult to explain in particular cases can become totally clear when the general situation is examined. If you are not too familiar with the algebra involved, do not worry: we shall return to such things in a later chapter. It does indeed take some justification, and this lends some merit to my student's wonder and to the worker's vexation. Even simple mathematics such as this can carry some degree of sophistication.

5. Whose performance is better?

In the name of efficiency and fairness, performance indicators are everywhere. Most of us are subject to them. One underlying pattern that normally emerges as cycles of performance measures are repeated is that performances tend to improve—and improve rather more than can be believed. This applies to everything from school pupils' examination results, to crime clean-up rates, to university research ratings.

The performance measures do concentrate the minds of those subject to them, but not so much on their performance as on achieving a good performance rating—people learn how the game works. The task of *measuring* performance is not as easy as you might expect. Even in the field of sport, difficulties emerge in quite innocuous situations. Here follows a very simple example.

The main performance indicator of a bowler in the game of cricket (for those more familiar with baseball, think of a bowler as a pitcher) is the average number of runs he concedes per wicket he takes—the lower the better. Suppose in one match one team has two bowlers, A and B, who return the following figures:

First innings: A takes 3 wickets for 60 runs while B takes 2 for 68.

Second innings: A takes 1 for 8 and B takes 6 for 60.

In the first innings A has the superior performance as he averages 20 runs per wicket while B's average is 34. In the second innings it is again A who has the better figures, for his average is 8 while that of B is 10. However, if we now look at the overall match performances of the two players we see that A took a total of 4 wickets for 68—an average of 17, while B took 8 wickets for 128—an average of 16. We thus are left with the unpalatable conclusion that B has a superior match performance to A, but A's performance, *using the same performance indicator*, is superior to that of B in both innings!

Our sixth problem is entirely different in nature: it is to verify a property of numbers that people often find quite striking.

6. Why does adding successive odd numbers always yield a perfect square?

$$1 = 1^2, 1 + 3 = 4 = 2^2, 1 + 3 + 5 = 9 = 3^2,$$
$$1 + 3 + 5 + 7 = 16 = 4^2, \ldots$$

Try one or two more cases. Indeed, you may feel up to writing a general formula expressing this conjecture: the sum of the first n odd numbers is n^2. You need to see how to express the nth odd number in terms of n in order to do this. Now the first odd number is 1, the second is 3, the third is 5, and so on—the pattern is to double the number and subtract 1, so that the nth odd number is $2n - 1$. ($2n$ is shorthand for $2 \times n$.) Therefore our conjecture may be written as:

$$1 + 3 + 5 + \ldots + (2n - 1) = n^2. \tag{1}$$

We have already checked (1) for the first four values of n, but can we come up with a convincing argument in general? There are quite a few. I will give one with a geometric flavour. The idea is to take a simple shape and to calculate its area in two different ways corresponding to either side of the equation. The obvious shape to try is a square of side length n, for its area is surely n^2. We then partition the square into non-overlapping bracket shapes as shown in Figure 3, the corner bracket being not a true bracket but a 1×1 square. Since each bracket is formed from the one it rests upon by adjoining two squares, one to either end, we see that the total area of these brackets is indeed $1 + 3 + 5 + \ldots + (2n - 1)$, as expected.

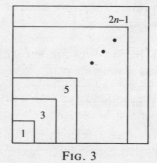

FIG. 3

As I have already mentioned, such jigsaw puzzle arguments can be used to prove many important results, including the celebrated theorem of Pythagoras, as will be seen in Chapter 3.

Our next problem is very similar, though the solution I have chosen is quite different.

7. What is the sum of the first n counting numbers?

We shall show that the answer is:

$$1 + 2 + \ldots + n = \frac{n(n+1)}{2}.$$

It is easy enough for you to check this works for small numbers; for instance, $1 + 2 + \ldots + 10 = 55 = \frac{(10 \times 11)}{2}$ as the formula suggests. The proof given here is a consequence of rearranging the sum. We shall pair the first and last terms, the second and second last terms, etc., together. This gives:

$$(1 + n) + (2 + n - 1) + (3 + n - 2) + \ldots$$

The point of this is that each pair of terms adds up to the same number: $n + 1$. All we have to do it seems is to multiply this by the number of pairs to get the answer. This is easy if n is itself an even number—the number of pairs is then $\frac{n}{2}$, and so we conclude that the total is $(n + 1) \times \left(\frac{n}{2}\right)$, which is the same as the expression given above: $\frac{n(n+1)}{2}$. For example, for $n = 10$ we are saying that:

$$
\begin{aligned}
1 + 2 + \ldots + 10 &= (1 + 10) + (2 + 9) + (3 + 8) + (4 + 7) + (5 + 6) \\
&= 11 + 11 + 11 + 11 + 11 \\
&= 11 \times 5 = 55.
\end{aligned}
$$

If n is odd, however, there is a little snag. It is impossible to break an odd number of things into pairs—the same approach will leave us with one number in the middle which has to be added separately. We can get around this with a little trick. We simply write a 0 in front of the whole sum. This does not change the answer, of course, but it does give us an even number of terms to add together once again, allowing us to recycle our original idea.

For example, for $n = 11$ we consider:

$$0 + 1 + 2 + \ldots + 10 + 11 = (0 + 11) + (1 + 10) + (2 + 9)$$
$$+ (3 + 8) + (4 + 7) + (5 + 6)$$
$$= 11 \times 6 = 66.$$

In general, for odd n, the argument runs like this. Put 0 in front of the sum and pair off as before. This time each pair sums to n, and the number of pairs is $\frac{(n+1)}{2}$, as there are $n + 1$ numbers in the whole sum because the 0 has been slipped on the front. Hence the total will be $\frac{n(n+1)}{2}$, just as in the even case.

This formula is important, as once it has been established it is a simple matter to find a formula for summing any so-called arithmetic series; but we shall wait until Chapter 7 before taking up this point again.

Our next problem is simple, but deceptive, and for that reason amusing. Imagine that we have a cable which runs right around the earth's equator (which we take to be a circle). It is decided that the entire cable has to be raised one metre above the surface.

8. By about how much must an equatorial cable be extended in order that it run one metre above ground?

Let us suggest four possibilities:

> A: 6 metres   B: 6 km   C: 600 km   D: 60,000 km.

Have you made your guess? The correct answer is A. Surprised or not, how do we arrive at the answer? We seem not to have enough information to solve the problem, for surely we need to know the circumference of the globe in order to calculate the answer. Maybe, maybe not. We proceed undaunted, using a symbol for the unknown: let r stand for the radius of the earth, so that the circumference is $2\pi r$ (remembering that π is about 3.14). That is to say, the original length of the cable is $2\pi r$. When we raise the cable one metre above the surface, the cable is now the circumference of a circle of radius $r + 1$, which has length $2\pi(r + 1)$. Now, all we want to know is the *difference* between these two circumferences, and we can at least write down an expression for this:

$$2\pi(r + 1) - 2\pi r.$$

Multiplying out the brackets (remember that, for any number a, $a(r + 1) = ar + a$, and 2π is just a number), we obtain:

$$2\pi r + 2\pi - 2\pi r = 2\pi,$$

and of course 2π is a little more than 6, showing that A is indeed the right choice.

Whether or not you find the answer amusing, the fact that we could answer the question at all is surprising. *We did not need to know the radius of the earth.* This has quite shocking consequences. Since the answer does not depend on the value of r, it will remain the same for any sphere, be it a basketball or even the planet Jupiter!

The fact that we assumed the equator to be a perfect circle does not affect our conclusions in any important way. The circular assumption allowed us to write the exact expression of $2\pi r$ for the circumference. Changing to another shape, even quite an irregular one, will alter this constant of proportionality a little, but a small number like answer A will continue to apply. More importantly, the independence of the answer from the size of the figure is still valid—for two similar shapes, be they circles, ovals, or more irregular figures, the increase in length of the cable is independent of the size of the planet in question. (Try the problem yourself for a cubic planet—you will discover that you now need 8 metres more cable.)

9. How can *n* men share a bottle of vodka?

I have been assured by a number of Russian colleagues that this is a very serious problem indeed. There is one bottle to share among *n* drinkers and each needs to be satisfied that he has his fair share. How can this be arranged?

With two, it is simple. One pours out into two glasses what he judges to be equal shares of the precious liquid, meaning that he who pours is happy to have either. The other gets to choose which glass is to be his. Neither can then complain.

It is not too difficult with a larger number to devise a workable scheme for any number *n*. The first, A, pours out what he claims to be a fair share. If any of the others think it is too much, one of their number, B say, can take A's glass and reduce its contents to what he deems to be the correct amount (without drinking it of course). It is assumed that no one will object if they believe that A is satisfied with what looks to be less than his share.

If anyone thinks B still has too much, that drinker can take B's glass and pour back what he thinks the excess. They continue this process. It is important to note that as the glass passes from hand to hand, none of the previous holders can object to the current level: for example A cannot claim that B has too much as B has less than what A has already said was a fair amount. At each stage the number of potential objectors decreases, so eventually we reach the situation where one of the party, X say, is holding a glass which he claims represents his fair share and none of the others are inclined to argue with him.

Mr X is now happy and retires from the procedure to have his drink. The others repeat the whole process with one less drinker and the remaining vodka until everyone has his drink and is happy.

Not everyone may be entirely happy though. This exhaustive system, apart from trying the patience of the participants, fails to ensure that no one will envy another's glass. It is true that no one can claim that he has not got his fair share, but one of those who settled carly in the process (like our Mr X who retired first) might be convinced that some later player got more than his share because the others were foolish enough to let him get away with it.

The mathematics involved in this question is more in the style of argument than in any clever calculation. This approach, where we reduce the case involving *n* to the case involving *n* − 1, is called an *inductive argument*.

Our next problem is also solved in this step-by-step fashion.

10. How long does it take to build the Tower of Hanoi?

The classical Tower of Hanoi problem consists of three pegs, A, B, and C, together with a tapering tower of concentric rings sitting on the first peg as shown in Figure 4. The task is to shift the tower from A to C subject to the following two constraints on the way you can move rings between the pegs:

(*a*) You may shift only one ring at a time.
(*b*) No ring may be placed on top of a smaller ring.

Try playing the game with a little tower of three or four coins—you will soon get the hang of how to do it. You have to find the *least* number of moves to complete the task.

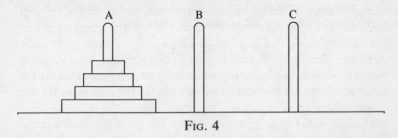

FIG. 4

Next, we need to find the minimum number of moves required for the *n*-ring game.

The mathematical feature that must be grasped is that, in order to play the *n*-ring game, you must first, in effect, play the $(n - 1)$-ring game. For instance, look at your 4-ring game, which is fully representative of the general situation. You cannot move the bottom big ring at all until you have shifted the tower of three rings above it on to another peg—that is, first you play the 3-ring game. Having done this, you can shift the big ring on to the required place in one move, whereupon there is no point in ever moving it again. To complete the procedure, you now must move the tower of three on to the largest ring—that is, you need to play the 3-ring game once more.

If we write a_4 for the *minimum* number of moves required to shift the tower of four rings, and let a_3 stand for the least number of moves needed for the 3-ring game, the above argument shows us that $a_4 = a_3 + 1 + a_3$, or in other words $a_4 = 1 + 2a_3$. Of course this argument works just as well for the *n*-ring game, telling us that, for any number $n = 2, 3, \ldots$,

$$a_n = 1 + 2a_{n-1},$$

where a_n stands for the least number of moves required to play the n-ring game to completion. Since it is clear that $a_1 = 1$ (one move does the 1-ring game), we can use this formula—the technical name is *recursion*—to calculate successive values of a_n; for instance, $a_2 = 1 + 2 \times 1 = 3$, $a_3 = 1 + 2 \times 3 = 7$, $a_4 = 1 + 2 \times 7 = 15$, ... The sequence of numbers we obtain is:

$$1, 3, 7, 15, 31, 63, 127, \ldots$$

Is a pattern emerging? Yes, if we start with the number 2 and double each time, we get almost the same sequence:

$$2, 4, 8, 16, 32, 64, 128, \ldots$$

The nth term in this latter sequence is 2^n, so it follows that a_n, the minimum number of moves needed to build the Tower of Hanoi, is given by:

$$a_n = 2^n - 1, \text{ for all } n = 1, 2, \ldots$$

The story that usually accompanies the Tower of Hanoi (which had 64 rings) is that the monks could move only one ring per day and when they had completed their task some great cataclysm would engulf all. We need not worry about this one. Even at 100 moves per day, the 20-ring game takes nearly 30 years—the monks' 64-ring game will take them billions of years.

That is not the end of the mathematical story, though. We can of course generalize the question and ask, what if we have 4 pegs? and generally k pegs? These questions yield similar but more complex problems. There is different mathematics to be found if we care to look at the actual sequence of moves and code them in a natural way as was pointed out in Martin Gardner's classic collection *Mathematical Puzzles and Diversions*. If we name the four rings, in order of increasing size, a, b, c, and d and play the 4-ring game, writing down the name of each ring as it is moved, we shall obtain the sequence:

$$a, b, a, c, a, b, a, d, a, b, a, c, a, b, a.$$

This sequence type is well worth noting as it arises in places other than this problem. For instance, take an old-fashioned ruler in which inches are divided in the binary fashion into halves, quarters, eighths, and sixteenths (Figure 5). If you read the markings from left to right, with the smallest (sixteenths)

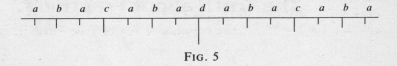

FIG. 5

corresponding to smallest ring a, the eighths corresponding to b, and so on, you read off just the same list as above. A mathematical phenomenon such as this one sometimes provides a common factor in two otherwise unrelated situations, which is often the key to understanding.

The Truth about Fractions

Mathematicians are not as fond of calculators as you might expect. Computers and calculators were of course invented and developed by mathematicians and engineers and are extremely useful, so why are we, at best, ambivalent towards them? The reason is that calculators are not much use when it comes to understanding the nuts and bolts of arithmetic, and they can act to replace thinking rather than stimulate it. Their widespread use in mathematics lessons can undermine learning. This fact has now been recognized in education and their indiscriminate use has been curtailed.

The other unfortunate effect of calculator usage is that it can serve to make the subject dull. Secondary school mathematics often is reduced to a button-pressing sequence which encourages the student to forget the maths, or at least to keep it at arm's length. The intellectual stimulation this offers is on a par with working a supermarket checkout. A hands-on approach is fine so long as it does not lead to brains-off! Typically, a student using a calculator writes little or nothing down, the effect of which is to render him or her mathematically inarticulate and incapable of doing a problem that involves more than one step.

In this chapter, however, I hope to exploit what advantages there are in using calculators. In doing so we shall meet the most unusual idea in this book, that of a so-called *uncountable* set. Its strangeness lies in its being furthest from the everyday world, although our starting point will be the familiar string of digits seen in a calculator display.

Calculators have allowed people to become more comfortable

with decimal displays, perhaps to an undesirable extent—frequently an ugly decimal approximation is preferred to a simple and accurate fraction. For example, how often do we see 66.7% written when the exact proportion is $\frac{2}{3}$?

Why is it good to be able to do arithmetic?

Ordinary arithmetic is quite difficult—it took mankind thousands of years to master it. A thorough understanding of the arithmetic of fractions takes much effort to acquire. Quite fundamental aspects of fractions were still being discovered in the nineteenth century. The so-called nth Farey series is simply the list of all fractions between 0 and 1 whose denominators do not exceed n, written in ascending order. For example, the fifth Farey series is:

$$\frac{0}{1}, \frac{1}{5}, \frac{1}{4}, \frac{1}{3}, \frac{2}{5}, \frac{1}{2}, \frac{3}{5}, \frac{2}{3}, \frac{3}{4}, \frac{4}{5}, \frac{1}{1}.$$

Farey sequences are replete with elegant algebraic and even geometrical properties and their discoverer was an amateur mathematician. That such an interesting and elementary aspect of mathematics had been missed by the collective body of mathematicians down the ages must have come as something of a shock to the great minds of the day, although it seems that basic properties of Farey series were first published as early as 1802 by Haros, who anticipated Farey's publication by some fourteen years.

Returning to the beginning, we have all heard children who are trying (perhaps not very hard) to do their sums say,

'What do I need to learn this for? If I ever wanted to know the answer I could just use my calculator.'

This kind of question is more often born of frustration, and those who ask it would not welcome a detailed answer. Not being able to deal with numbers can be a constant problem. If we cannot do our sums we are forced to dive for cover whenever numerical matters arise. Even the advent of calculators does not much help a seriously innumerate person, who can never feel confident that he or she has used the machine properly—it is of little more use than is a dictionary to someone illiterate.

Dealing with ordinary number and measurement questions requires training to a level that is somewhat beyond what you are likely to need in practice. This is because the problems you meet have to be well within your grasp in order for you to be able to deal with them confidently in a practical situation.

Do we need to know our tables? Yes, and I will explain why. The familiarity with the number system that the learning of tables engenders is worthwhile in itself, but there is a fundamental mathematical aspect to the situation as well. The point to appreciate is that the multiplication tables represent not a collection of arbitrary facts, like a list of telephone numbers, but rather the minimum set of products you need to know in order to do ordinary arithmetic.

Let us look at something even more basic—addition. In order to be able to do additions we need to know our addition tables up to 10; for example, when adding $59 + 17$ we need to know what $9 + 7$ is. In the same way, we need to memorize our multiplication tables up to 10 in order to learn how to multiply two numbers together. (Incidentally, the numbers $a + b$ and $a \times b$ are known as the *sum* and *product* respectively of a and b; the number $\frac{a}{b}$ is called the *quotient* of a over b.) If we do not know these additions and multiplications by heart, we shall be forced to rediscover them every time.

Why is this particular collection of facts necessary in order to do arithmetic? To be sure, there is some arbitrariness here, but it was introduced right at the beginning of the development of arithmetic when we decided to work in base 10. We were free to make this choice, but now we pay for it through the need to learn our addition and multiplication tables up to 10—if we had decided to work in binary, we would have only trivial additions and multiplications to memorize.

Multiplication tables have been traditionally taught up to 12. This is because many measurement systems were based on 12 (pounds, shillings, and pence, feet and inches, etc.) and since products up to 12×12 cropped up very frequently they were worth committing to memory. They still are, although the argument for doing so has become a little less compelling.

In order to understand numbers to a useful extent, a pupil needs to do lots of arithmetic. It is not the answers that are important, but the development of the skill required to obtain them. Doing arithmetic instils a basic familiarity with numbers and a confidence in handling them. All higher mathematics involves the same kinds of manipulation, performed with algebraic symbols instead of particular numbers, so a student needs to be completely sure-footed arithmetically in order to make these manipulations second nature. Lack of mastery of the basics leaves an obstacle to all future understanding of new material. In particular, we must be able to deal with fractions in order to have any real mathematical facility.

This book is not a refresher course on such things, but I will take the opportunity to say something on the matter. Should you be totally at ease with the arithmetic of fractions, I would still invite you to read the remainder of this section—reading things that one already knows can be quite enjoyable and I still hope to present a surprise or two.

The arithmetic of fractions requires a clear idea of equality of fractions. If you imagine a pie sliced into halves and then into quarters you see that, although the notions of $\frac{2}{4}$ and $\frac{1}{2}$ are different, they none the less represent equal portions of pie. The fractions $\frac{1}{2}$, $\frac{2}{4}$, $\frac{3}{6}$, etc. are equivalent. We shall say that they are *equal*, although this is inaccurate unless explained: $\frac{1}{2}$ and $\frac{2}{4}$ are *different* fractions, but they are equal in that they represent equal quantities. The other fortunate aspect of the situation is that, although there are any number of fractions equal to a given one, only one of them is *reduced*; this means that it has been cancelled to lowest terms in that the top line, the *numerator*, and the bottom line, the *denominator*, have no common factor other than 1. For example, the fractions $\frac{6}{15}$ and $\frac{12}{30}$ each cancel to $\frac{2}{5}$, so that $\frac{2}{5}$ is the reduced form of both fractions. The reduced version is naturally pre-eminent—it is the simplest as it has a smaller numerator and denominator than all the rest.

Equality of fractions can also be expressed through cross-multiplication:

$$\frac{a}{b} = \frac{c}{d} \Leftrightarrow ad = bc.$$

(The double-headed arrow reads 'is equivalent to'; the single-headed version $\Rightarrow$ denotes 'implies'.)

Indeed, more generally we can test whether or not one positive fraction is less than or equal to another through cross-multiplication:

$$\frac{a}{b} \le \frac{c}{d} \Leftrightarrow ad \le bc.$$

(As a mnemonic, inequality symbols such as $\le$, less than or equal to, always point to the smaller of the two numbers.) The rule for comparing fractions also works with $\le$ replaced by any of $<, >,$ or $\ge$. The rule is valid because inequalities are respected when both sides are multiplied by positive numbers and we can pass from the first inequality above to the second through multiplying both sides by bd. For example:

$$\frac{2}{3} < \frac{5}{7} \Leftrightarrow 14 < 15.$$

We are now in a position to give a general rule for the addition or subtraction of two fractions $\frac{a}{b}$ and $\frac{c}{d}$. We first replace the fractions by two equal fractions having a *common denominator*. A common denominator can always be found by multiplying the denominators together yielding bd. Since $\frac{a}{b} = \frac{ad}{bd}$ and $\frac{c}{d} = \frac{bc}{bd}$, we obtain:

$$\frac{a}{b} \pm \frac{c}{d} = \frac{ad}{bd} \pm \frac{bc}{bd} = \frac{ad \pm bc}{bd}. \tag{1}$$

The sign $\pm$ stands for 'plus or minus', and it is used in order to kill two birds with one stone. This rule always works but the answer that results may not be reduced even though the original fractions were. For example:

$$\frac{1}{6} + \frac{2}{9} = \frac{1 \times 9 + 2 \times 6}{6 \times 9} = \frac{9 + 12}{54} = \frac{21}{54} = \frac{7}{18}.$$

It is good to have a rule like (1) which gives the answer every time, but there are drawbacks. In principle, (1) contains all the information you need in order to add and subtract fractions, but it would leave someone unfamiliar with the subject cold as it does not convey the underlying idea of the operation. It should be regarded more as a summary of what is going on. Second, the rule does not always represent the best way of going about a particular sum. In practice, it is best to seek the lowest common denominator; this is the *least* multiple of b and d. The product bd is a multiple of b and d but not necessarily the least one. In general this least multiple is given by $\frac{bd}{h}$, where h is the highest common factor of b and d, and we shall have more to say on this matter in Chapter 4. In our example above the value of h is 3, so that the least common denominator is $\frac{(6 \times 9)}{3} = 6 \times 3 = 18$. We then have:

$$\frac{1}{6} + \frac{2}{9} = \frac{3}{18} + \frac{4}{18} = \frac{7}{18}.$$

Multiplication of fractions is easier than addition: we simply multiply numerators and denominators. Once again, however, the answer so obtained may not be reduced:

$$\frac{5}{12} \times \frac{9}{10} = \frac{45}{120} = \frac{3}{8}.$$

It is important to have the presence of mind to look for factors *before* you carry out the multiplications, because it is then that the possible cancellations are most apparent:

$$\frac{\cancel{5}^{1}}{\cancel{12}^{4}} \times \frac{\cancel{9}^{3}}{\cancel{10}^{2}} = \frac{1}{4} \times \frac{3}{2} = \frac{3}{8}.$$

There is a general point to be made here which students of mathematics often take a long time to realize. It was easier to cancel the fraction $\frac{45}{120}$ when it was written as a product $\frac{5}{12} \times \frac{9}{10}$. It is often simpler to perform arithmetic on a number, or algebra on an algebraic expression, when it is in the form of a product than when it has been 'multiplied out'.

Unfortunately, a student anxious to get the answer can often overlook this and so perform unnecessary multiplications which are counter-productive. I am afraid that with a calculator at hand the temptation can prove irresistible. Having got the correct answer, it takes rare presence of mind on the student's part to analyse what he or she has done and eliminate the unnecessary steps. This is where a good teacher can help.

It is straightforward to see that our rule for fraction multiplication makes sense. If a pie is sliced into b equal parts and then each of those slices is cut into d equal parts, we have divided the pie into bd equal pieces; that is to say:

$$\frac{1}{b} \times \frac{1}{d} = \frac{1}{bd}.$$

Multiplying this by the numerators a and c gives the general rule:

$$\frac{a}{b} \times \frac{c}{d} = \frac{ac}{bd}.$$

Finally, to divide a pie by n means to take $\frac{1}{n}$th part of it. In general, to divide by $\frac{a}{b}$ is to multiply by the *reciprocal* $\frac{b}{a}$. In summary:

$$\frac{a}{b} \div \frac{c}{d} = \frac{a}{b} \times \frac{d}{c} = \frac{ad}{bc}.$$

This truly makes division the inverse operation to multiplication, for if we multiply by $\frac{a}{b}$ and then divide by it, the combined effect of both operations is to multiply by $\frac{a}{b} \times \frac{b}{a} = \frac{ab}{ba} = 1$. None the less, division of fractions is often regarded as a mystery. It is not that people cannot do the sums, it is just that the rule 'invert the second fraction and multiply' manages to remain obscure. The best way to make the process convincing is to do a division directly and observe that the net effect of what you have done is described by the preceding rule.

For example, what is $\frac{2}{3} \div \frac{3}{4}$? We have:

$$\frac{2}{3} \bigg/ \frac{3}{4}.$$

Let us rid ourselves of the denominator on the bottom by multiplying each fraction by 4: the overall effect of this is to multiply by $\frac{4}{4} = 1$ so the value of the fraction is unchanged:

$$\left(\frac{2}{3} \times 4\right) \bigg/ \left(\frac{3}{4} \times 4\right) = \left(\frac{2}{3} \times 4\right) \bigg/ 3 = \frac{2}{3} \times \frac{4}{1} \times \frac{1}{3} = \frac{2}{3} \times \frac{4}{3} = \frac{8}{9}.$$

Hence to divide by $\frac{3}{4}$ is the same as multiplying by $\frac{4}{3}$.

The use of fractions can be seen in the records of the ancient Egyptians who used unit fractions such as $\frac{1}{5}$ and $\frac{1}{20}$ freely but were reluctant to regard fractions such as $\frac{2}{7}$ as having the same status, although they had a special symbol for $\frac{2}{3}$. They would for example express $\frac{2}{7}$ as $\frac{1}{4} + \frac{1}{28}$. (What seems the obvious alternative to us, $\frac{1}{7} + \frac{1}{7}$, did not appeal to them.) This, however, leads to a real problem. Is it possible to write any proper fraction, that is a fraction between 0 and 1, as a sum of distinct unit fractions? The answer is 'yes', and one way of going about it will afford you the opportunity of brushing up on your arithmetic. Begin with the given fraction, $\frac{m}{n}$, and subtract the largest unit fraction you can. Do the same to the remainder and keep repeating the process. This will yield a required decomposition. For example take $\frac{9}{20}$. Subtracting $\frac{1}{3}$ leaves $\frac{7}{60}$ and subtracting a further $\frac{1}{9}$ gives $\frac{1}{180}$, yielding the 'Egyptian' decomposition:

$$\frac{9}{20} = \frac{1}{3} + \frac{1}{9} + \frac{1}{180}.$$

This greedy approach of always subtracting the largest reciprocal available does work but may not always yield the shortest sequence of unit fractions possible, as we can see even in this example, because $\frac{9}{20} = \frac{1}{4} + \frac{1}{5}$.

Try the method yourself on $\frac{5}{7}$ and on $\frac{6}{13}$ —you should be able to write each as the sum of three unit fractions.

A whole host of questions is suggested by this ancient problem, many of which are still being wrestled with by mathematicians today. The simplest one is: how do we find the largest reciprocal less than a given fraction? I shall answer this in Chapter 5 along with the fundamental question: how do we know this process works? In order for the process to stop we must eventually reach the stage where the remainder is itself a unit fraction. It is conceivable that this never happens and we may go on subtracting reciprocals for ever. Rest assured this is not the case, and we shall see in Chapter 5 that the proper fraction $\frac{m}{n}$ can always be written as the sum of m or fewer distinct unit fractions.

What goes on in decimal arithmetic?

Expressions involving fractions with many different denominators are a nuisance. We can deal with a multiplicity of denominators by finding a common denominator for all the fractions involved. This will allow us to deal with any particular problem, but it would be nice if there were a single common denominator for all fractions. Of course there isn't. We can attempt to circumvent this difficulty by recourse to decimal expansions. This allows us to display all fractions in a uniform manner—the price we pay, however, is that our representation of fractions, even very simple ones, generally becomes infinite.

Almost everyone knows that $\frac{1}{3} = 0.33333...$ The left-hand side of this equation represents a simple idea, that of an ordinary fraction, while that on the right involves an infinite process—one that goes on for ever. If this does not worry you, multiply both sides by 3: you get $1 = 0.99999...$ There is nothing wrong here, but I find that people often do not like the look of this and immediately begin to protest, insisting that the right-hand side is somehow less than 1. How much less? I ask. The response to this awkward question is sometimes the suggestion that $0.99999...$ represents the number that immediately precedes the number 1 and that the two are separated only by an infinitesimal gap. This sounds more scientific, but there is no such number—no number *immediately* precedes 1. We do, however, have to face up to something that you may not have noticed before—that one number can have two different decimal expansions. None the less, this is little more than a nuisance and there is essentially only one type of exception—a terminating decimal such as 2.364 is equal to $2.36399999...$ also. That said, let us examine the link between ordinary fractions and their decimal expansions. Until further notice, we shall be considering only positive numbers— the use of negative numbers is important of course, and we shall talk of them later, but they do not have anything to contribute to the problem of decimal representations and so they need not concern us for the time being.

There are two types of ordinary or *vulgar* fraction: proper and improper. A *proper* fraction is one in which the numerator is less than the denominator, e.g. $\frac{2}{3}$, $\frac{8}{17}$, etc. All such fractions represent numbers between 0 and 1. A top-heavy fraction, such as $\frac{25}{12}$, is called *improper*. By dividing the denominator into the numerator on such occasions we can express such a fraction as a *mixed number*, in this case $2\frac{1}{12}$, which consists of a whole number followed by a proper fraction. The mixed-number form of fraction is awkward to use in calculation and so the improper fraction representation should be preferred. However, it is often best to write the final answer to a sum as a mixed number as it more clearly reveals its size; for example, writing $\frac{47}{7}$ as $6\frac{5}{7}$ tells you at sight that you are dealing with a quantity between 6 and 7.

If we understood all about the decimal expansions of numbers between 0 and 1, we would understand decimal expansions generally, so let us concentrate on the interval from 0 to 1.

A *rational* number is one that can be written as a fraction or, as we sometimes say, as a ratio of two whole numbers. As you know, two different fractions can represent the same number, $\frac{1}{2}$ and $\frac{2}{4}$ for instance. Here again we have the same number written in different ways, so that it is not only in decimal expansions that we meet this type of inconvenience. By cancelling to lowest terms, however, we can represent any rational as a fraction of the form $\frac{a}{b}$, where a and b have no common factor other than 1. We can think of the rationals therefore as the set of all fractions that have been cancelled down in this way.

What happens when we write a rational as a decimal fraction? The answer is that we always obtain a *recurring* decimal, that is to say a decimal where there is a block of digits that repeats itself indefinitely after a certain point in the expansion. We indicate this by writing dots over the first and last digits of the block. Some examples:

$$\frac{2}{3} = 0.666... = 0.\dot{6}, \ \frac{1}{7} = 0.142857142857... = 0.\dot{1}4285\dot{7},$$

$$\frac{1}{24} = 0.041666... = 0.041\dot{6}, \ \frac{1}{17} = 0.\dot{0}58823529411764\dot{7}.$$

You may think that I have forgotten some of your old friends like $\frac{1}{2} = 0.5$ and $\frac{3}{8} = 0.375$—the terminating decimals. Not really: terminating decimals such as these are just special recurring decimals, by which I mean to say that $\frac{1}{2} = 0.5\dot{0}$ and $\frac{3}{8} = 0.3750\dot{0}$, but of course there is usually no need to write them in this way.

There are several questions to be answered.

1. Why do rationals lead to recurring decimals as I claimed just now?
2. Which rationals lead to terminating decimals?
3. What can be said about the length of the repeating block in the expansion? (In the four examples above the lengths of the repeating blocks were respectively 1, 6, 1, and 16.)
4. Can every recurring decimal be converted back to a fraction? If so, how?

To see why fractions lead to recurring decimals, it is best to look again at the way you were taught to turn a fraction such as $\frac{5}{6}$ into a decimal:

$$\frac{5}{6} = 0.8333... = 0.8\dot{3}.$$

The way you would have been taught to do it is to say;

1. 6 into 5 does not go, so write down 0. (indicating that the fraction is less than 1) and carry the 5;
2. 6 into 50 goes 8 with 2 remainder; write down the 8 and carry the 2.

What you are doing here is to treat the 5 as $50 \times \frac{1}{10}$; we can divide 6 into 50 giving 8 (meaning $\frac{8}{10}$ of course) and leaving a remainder of 2, representing $\frac{2}{10}$, which is still to be divided by 6. We treat the $\frac{2}{10}$ as $20 \times \frac{1}{100}$ in the next step in the division:

3. 6 into 20 goes 3 times with 2 remainder; write down the 3 and carry the 2.

At this stage you have shown that $\frac{5}{6} = 0.83 + (\frac{2}{100} \div 6)$, and we continue working on this remainder in the same way. Of course, in this case the remainder is never 0 so the process goes on for ever. However, since all the remainders are equal to 2 from this point on, we are fated to repeat the same little calculation over and over, yielding:

$$\frac{5}{6} = 0 \cdot 8\dot{3}.$$

We can now answer our first question. When converting the fraction $\frac{m}{n}$ to a decimal, it will either terminate or not. If it does not, the remainder after each stage in the division must always be one of the numbers 1, 2, ..., $n-1$. Since there are just $n-1$ possibilities, a remainder *must be repeated somewhere in the first n steps.* As soon as a remainder arises for the second time, we will be forced to repeat exactly the same cycle of remainders that we have just been through. This cycle, of course, ends with the same remainder being repeated a second time, and so we are then caught in this loop for ever.

For example, $\frac{1}{7}$ is a non-terminating decimal: the possible remainders met when doing the division are the numbers 1 through to 6, and indeed they do all arise. On doing the division of 1 divided by 7, the cycle of remainders is 1, 3, 2, 6, 4, 5, 1, 3, 2, 6, 4, 5, 1, ... and so the length of the repeating block is six.

This answers the first question and goes some way to answering the third: what is the length of the repeating block? If the denominator is n, the block length is at most $n-1$. This maximum possible length is sometimes attained—fractions with denominators of 7 or 17 have blocks of lengths 6 and 16 respectively, as we have already seen. Murphy's Law does not apply, however, in that things are not always as bad as possible, even for prime denominators: $\frac{1}{11} = 0.\dot{0}\dot{9}$, a block length of only two, and $\frac{1}{13} = 0.\dot{0}7692\dot{3}$, with a block length of six. More can be said about the length r of the block in the decimal expansion of $\frac{m}{n}$. So long as m and n have no common factor, r depends only on n and not on m. The value of r itself can be described in other ways, but not as conveniently as you might hope: there is no quick general rule for finding r from the the value of n.

On the other hand, the second question of recognizing what fractions yield terminating decimals is more easily dealt with. We all know that $\frac{1}{2} = 0.5$ and $\frac{1}{5} = 0.2$, and 2 and 5 are factors of 10, the base of our number system. Now if we take two terminating decimals, we can multiply them together and the result will be another terminating decimal. You will remember that if the first number has r decimal places and the second has s, then their product has no more than $r + s$ places; for example, $0.202 \times 0.01744 = 0.00352288$ is an instance where $r + s = 3 + 5 = 8$, and the product terminates eight places after the decimal point. It follows that any number which is a product of $\frac{1}{2}$ and $\frac{1}{5}$ any number of times, that is to say whose denominator is a product of 2s and 5s, will have a terminating expansion. For example:

$$40 = 2^3 \times 5 \text{ and } \tfrac{1}{40} = 0.025; \; 16 = 2^4 \text{ and } \tfrac{1}{16} = 0.0625.$$

What is more, any multiple of a terminating decimal will also terminate; for example, $\frac{7}{40} = 0.175$. The reason for this is that multiplying a terminating decimal by a whole number will not increase the number of non-zero entries required after the decimal point (although it can decrease it; for example, $0.25 \times 2 = 0.5$).

It is even simpler to see that the reverse is true: a terminating decimal can be written as a fraction in which the denominator is a product of 2s and 5s, as any terminating decimal can immediately be written as a fraction whose denominator is a power of 10. For example:

$$0.255 = \frac{255}{1000} = \frac{51}{200}.$$

In this example the denominator is

$$1000 = 10 \times 10 \times 10 = 2 \times 5 \times 2 \times 5 \times 2 \times 5 = 2^3 \times 5^3.$$

Of course, it may be possible to cancel the fraction down further, as it is here, but the denominator will remain a product of the numbers 2 and 5 ($200 = 2^3 \times 5^2$).

We have arrived at a complete description of fractions that yield terminating decimals:

A fraction $\frac{m}{n}$ has a terminating decimal expansion if and only if n has the form $n = 2^a 5^b$, that is to say, if and only if the denominator of the fraction is a product of 2s and 5s. (This includes denominators that are products of only 2s or only 5s such as $\frac{1}{16}$ or $\frac{1}{25}$.)

The numbers 2 and 5 are special only in so far as they are factors of 10, which is the base in which we work. If we were to move our base, then the class of terminating decimals would move with it. For example, in base 3 (known as *ternary*) the fraction $\frac{1}{3}$ is terminating for in ternary its expansion is 0.1, the 1 now standing for $1 \times \frac{1}{3}$, and not for $1 \times \frac{1}{10}$.

We now turn to our fourth question. I shall show how to change any recurring decimal into a fraction. This particular technique seems not always to be taught in schools, which is a shame, as it is both simple and rather clever. A couple of examples should be enough to clarify the method.

Let us try $0.\dot{6}\dot{3}$. The length r of the block here is 2, so we multiply the number, let us call it a, by $10^2 = 100$. Now $0.\dot{6}\dot{3} = 0.6363$ and so this gives $100a = 63.\dot{6}\dot{3}$. The point of doing this is that, since both $100a$ and a have exactly the same expansion after the decimal point, it shows that $100a = 63 + a$. Subtracting a from both sides we obtain $99a = 63$, so that $a = \frac{63}{99}$. Finally, we cancel down to obtain:

$$0.\dot{6}\dot{3} = \frac{7}{11}.$$

It is best now to try a couple of these yourself. Use the same technique to check that $0.\dot{1}\dot{8} = \frac{2}{11}$, and $0.\dot{0}3\dot{7} = \frac{1}{27}$. (You need to multiply by 1000 here.)

A slight variation arises when we take an example such as $a = 0.2\dot{7}$. In this case our $r = 1$, so we need only multiply by 10 to get $10a = 2.\dot{7}$. Subtraction then yields $9a = 2.\dot{7} - 0.2\dot{7}$. This time the two numbers are identical from the *second* place after the decimal point onwards, so these parts cancel each other leaving $9a = 2.7 - 0.2 = 2.5$. Multiplying both sides by 10 in order to have an equation involving only whole numbers gives us $90a = 25$, and so $a = \frac{25}{90} = \frac{5}{18}$.

Another one to try: show that $0.58\dot{3} = \frac{7}{12}$.

In conclusion, we can represent any fraction as a recurring decimal (remembering that terminating decimals lie in this category) and vice versa, thus establishing a correspondence between the rational numbers and the recurring decimals. Of course it is easy enough to produce decimals that are not recurring. For instance, in the number

$$b = 0.101001000100001000001...,$$

there is a pattern to this expansion yet it is not a recurring decimal. We conclude that b is not a rational number—it cannot be written as the ratio of two whole numbers. Numbers like our b, known as *irrationals*, are very easy to find. For instance, can you see why 0.12345678910111213141516... is similarly another irrational?

Irrationality in geometry

Indeed, it is not hard to generate numbers on your calculator that have no apparent pattern to their decimal expansion at all. Try $\sqrt{2} = 1.414213...$ This one requires some thought. How do we know that $\sqrt{2}$ does not have a recurring decimal expansion? It might be that the recurring block is hundreds of digits long, or that the recurrence does not begin until after millions of decimal places. In other words, it may still be rational after all.

It is said that the Pythagoreans in the sixth century BC worried terribly about numbers like $\sqrt{2}$. They certainly would not have been happy with our way of doing things. After all, if we cannot write $\sqrt{2}$ as a fraction, what meaning does it have?

In our approach, through decimal expansions, our philosophical position amounts to the following. We say that a number is *real* if we can show that it has a decimal expansion. For that reason, $\sqrt{2}$ is real because we can find its expansion to any number of places as follows. We begin by noting that $1^2 < 2 < 2^2$, and so, taking square roots, we see that $1 < \sqrt{2} < 2$, which is to say that the number $\sqrt{2}$ lies between 1 and 2 and so $\sqrt{2} = 1. \cdots$ Next, we note that $1.4^2 = 1.96 < 2 < 2.25 = 1.5^2$, and so $1.4 < \sqrt{2} < 1.5$. We can continue in this way—to two and to three decimal places. You can verify that $1.41 < \sqrt{2} < 1.42$, $1.414 < \sqrt{2} < 1.415$, and so on. In principle there is no limit to the number of places to which we can calculate $\sqrt{2}$, and therefore to our way of thinking $\sqrt{2}$ is a real number, even though it may turn out that it is not rational. (To be sure, there are more efficient ways of extracting square roots than this naive approach, but it is quite enough to make the point.)

From what I have read, I believe that the Pythagoreans would have had none of this. They were believers in simplicity and had a deep distrust of any infinite process such as the one in which we have just indulged. They would not accept that something introduced through a process of endless calculation could enjoy the same status as the ordinary rational numbers which they so cherished and which formed the cornerstone of their philosophy. None the less, they believed in $\sqrt{2}$ as well, but for quite different reasons. To them $\sqrt{2}$ was a meaningful number because it could be constructed. To explain their outlook, we need to adopt a geometrical approach.

Pythagoras's Theorem is a simple fact about any right-angled triangle. If we let the shorter sides have lengths a and b and the hypotenuse have length c, then the theorem says that $a^2 + b^2 = c^2$. In particular, if we take $a = b = 1$ we obtain $c^2 = 1^2 + 1^2 = 1 + 1 = 2$. Therefore the length of the longest side of this triangle is $c = \sqrt{2}$. Now the Greeks knew that any such triangle could be constructed without use of a device to measure length or angle but simply through the use of a straight edge and compasses. In their eyes that meant that constructible numbers like $\sqrt{2}$ enjoyed a natural existence which they deemed to be especially important.

Let us see how $\sqrt{2}$ is constructed. To say that a number a is *constructible* means that, given any line segment to act as a standard of unit length, there is a sequence of operations that can be performed using a straight edge (not a marked ruler, just an edge) and compasses which leads to another line segment of length a. To construct the isosceles right-angled triangle we mentioned above, proceed as follows. ('Isosceles' means that two sides, and therefore two angles of the triangle, are equal.)

You are given a line segment with endpoints A and B to act as our standard unit length. Extend the line segment past B and use your compasses to mark a point C to the right of B so that AB and BC are of equal length (see Figure 1).

Open the compasses up further and draw a segment of a circle with centre at A and, without changing the radius of the compass setting, do the same at C. The two circles you have drawn will meet above and below B: let D be the intersection above B. Draw the line from B through to D. By symmetry, the angle ABD is a right angle. Again, use the compasses to mark a length equal to that of AB along the line through B and D. Mark this final point

E which, by construction, is the third vertex of our right-angled triangle with sides of length 1 (*AB*), 1 (*BE*), and, by Pythagoras, $\sqrt{2}$ (*AE*).

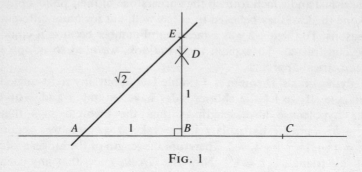

FIG. 1

The fact that $\sqrt{2}$ turned out to be irrational upset the way of thinking of the Pythagoreans. There are even stories about death pacts and murders carried out in order to suppress the dreadful news. This sounds ridiculous to our way of thinking, and since thousands of years separate the lives of these people and our own, such tales raise no more than a smile. Let us see why it is impossible that $\sqrt{2}$ could be equal to some ordinary fraction. The argument is by way of *contradiction*: we suppose otherwise and then go searching for trouble.

Suppose, contrary to what we wish to show, that $\sqrt{2}$ is a rational number $\frac{a}{b}$, where we have cancelled any common factor that *a* and *b* might have had. Squaring both sides of $\sqrt{2} = \frac{a}{b}$ gives $2 = \frac{a^2}{b^2}$, which is the same as saying

$$2b^2 = a^2.$$

Now the left-hand side is a multiple of 2; that is to say, it is an even number, and so a^2 is even also. It follows that *a* is itself even. (The product of any two odd numbers is odd, so if *a* were odd, so would be a^2.) We can therefore write *a* as 2*c* for some whole number *c*. This yields:

$$2b^2 = (2c) \times (2c) = 4c^2.$$

Cancelling the common 2 in this little equation gives $b^2 = 2c^2$.

Can you see the difficulty coming? Using the same reasoning as before, we now conclude that b, as well as a, must also be even. But this contradicts our original assumption that a and b had no common factor—the very idea that $\sqrt{2} = \frac{a}{b}$ has led to the conclusion that both a and b are multiples of 2. We are left with no choice but to admit that it is impossible to write $\sqrt{2}$ as a fraction. I shall show, however, in Chapter 9 that it is possible to express $\sqrt{2}$ as a recurring expansion of quite a different kind.

Modern mathematics delights in results of this kind. It just goes to show that there is more to the world than ratios of whole numbers. The Ancients passionately yearned for a system of philosophy which would encompass everything and so were bitterly disappointed by new discoveries that violated their beliefs. There are still those of us who hanker after a complete picture of the entire universe but this attitude is more likely to impede progress than to foster it. Time after time new aspects of science have flourished only when people have relaxed and pursued new ideas free of inhibitions, prejudice, or the need to justify what they were doing in terms of a philosophy, whether it be religious or secular.

Once you have identified one irrational, the floodgates open as you can immediately generate infinitely many. Suppose that x is irrational. (Take x to be $\sqrt{2}$ if you wish.) Then, for any rational number $\frac{a}{b}$, be it positive or negative, $x + \frac{a}{b}$ is also irrational, for if on the contrary it were equal to some rational $\frac{c}{d}$, we could conclude that:

$$x = \frac{c}{d} - \frac{a}{b} = \frac{(bc - ad)}{bd},$$

which is itself rational, contradicting the irrationality of x. A similar thing happens if we multiply x by the rational $\frac{a}{b}$. As long as a is not 0, this product cannot equal a rational number $\frac{c}{d}$ as this would yield again that x was equal to a rational (the centred dot stands for multiplication):

$$\frac{a}{b} \cdot x = \frac{c}{d} \Rightarrow x = \frac{b}{a} \cdot \frac{c}{d} = \frac{bc}{ad}.$$

In particular, the numbers $1 + \sqrt{2}$ (adding 1) and $\frac{\sqrt{2}}{3}$ (multiplying by $\frac{1}{3}$) are irrational in consequence of the irrationality of $\sqrt{2}$.

Something worthy of note is that it is perfectly possible to add or to multiply two positive irrationals together and arrive at a rational answer. For example, $2\sqrt{2}$ and $\sqrt{2}$ are positive irrational numbers, yet $2\sqrt{2} \times \sqrt{2} = 2 \times 2 = 4$. In the same way, $2 - \sqrt{2}$ and $\sqrt{2}$ are both irrational but their sum is 2.

More surprisingly, there is a neat argument that proves that there are two irrational numbers, a and b, such that a^b is rational. We are going to show this although we will not be able to find out what the numbers a and b actually are! First I will remind you about the way powers behave.

Indices, logs, and irrationality

The *First Law of Indices* is that $a^n \times a^m = a^{n+m}$. This is obvious once you note that the indices n and m merely count the number of factors in the product; so, for instance:

$$a^2 \times a^3 = (a \times a) \times (a \times a \times a),$$

which of course equals a multiplied by itself $2 + 3 = 5$ times. In a similar way we can make sense of the *Second Law of Indices* through cancellation; $\frac{a^n}{a^m} = a^{n-m}$, for example:

$$\frac{a^5}{a^2} = \frac{(a \times a \times a \times a \times a)}{(a \times a)} = a^{5-2} = a^3.$$

Finally the *Third Law of Indices* is by the same token a counting statement: $(a^n)^m = a^{nm}$; for example:

$$(a^2)^3 = (a \times a) \times (a \times a) \times (a \times a) = a^{2 \times 3} = a^6.$$

Meanings are given to powers that are not positive whole numbers by insisting that these rules continue to hold always. For instance, by $a^{\frac{1}{2}}$ we mean $\sqrt{a}$, for then:

$$a^{\frac{1}{2}} \times a^{\frac{1}{2}} = \sqrt{a} \times \sqrt{a} = a = a^1,$$

and so, with this interpretation,

$$a^{\frac{1}{2}} \times a^{\frac{1}{2}} = a^{\frac{1}{2}+\frac{1}{2}} = a^1 = a,$$

consistent with the First Law. By a^{-1} we mean $\frac{1}{a}$, as this is consistent with applying the Second Law in situations such as

$$\frac{a}{a^2} = \frac{1}{a};$$

subtracting the indices here leads to a power of $1 - 2 = -1$. The Second Law also requires that we take $a^0 = 1$ in order to comply with the fact that $\frac{a^2}{a^2} = 1$ as subtraction of the indices here leaves a power of $2 - 2 = 0$.

Given the Laws of Indices, we can show that there must exist irrationals a and b, such that a^b is rational. First, consider trying $a = b = \sqrt{2}$. Either $\sqrt{2}^{\sqrt{2}}$ is rational, or it is not. If it is, we are finished since this would be our example. If on the other hand $\sqrt{2}^{\sqrt{2}}$ is not rational (which you might expect is the more likely alternative), put $a = \sqrt{2}^{\sqrt{2}}$ and $b = \sqrt{2}$. Then both would be irrational; yet using the Third Law of Indices we obtain:

$$a^b = (\sqrt{2}^{\sqrt{2}})^{\sqrt{2}} = (\sqrt{2}^{\sqrt{2}\sqrt{2}}) = (\sqrt{2})^2 = 2.$$

It follows that, in either case, the required irrational numbers must exist.

The thing to spot in this proof is that it gives two alternatives and concludes that one of them leads to an example of a pair of numbers with the required property, but provides no clue as to which pair does the job. For this reason many people, including some mathematicians, consider such a proof practically worthless. It does not, however, trouble me.

Since we have taken the time to review the Laws of Indices, we have the opportunity to say a few words about logarithms, a topic every older reader will probably have dealt with at length during their schooldays.

The definition is simple: if $y = 10^x$, we say that x is the logarithm of y to the base 10 and write $x = \log_{10} y$. We can replace 10 by some other base b but we will have no call to do that here—we will use only base 10 and so write $x = \log y$ to mean $y = 10^x$; for example, $\log 1000 = 3$ as $10^3 = 1000$, $\log 0.1 = -1$ as 10^{-1} means $\frac{1}{10} = 0.1$.

The magical property of logarithms that so revolutionized science was that they turned multiplications and divisions into additions and subtractions, because:

$$\log ab = \log a + \log b; \quad \log\left(\frac{a}{b}\right) = \log a - \log b.$$

This allowed difficult multiplications and divisions to be done with tolerable accuracy: to multiply a and b, one only had to look up their logarithms, add them, and then find the number whose logarithm was equal to this sum—that is, look up the anti-logarithm. The above properties are simple consequences of the definition of logs and the Laws of Indices. For instance, the additive property stems from the First Law of Indices. Write x and y for $\log a$ and $\log b$ respectively. Then

$$a = 10^x, b = 10^y \quad \text{and so} \quad ab = 10^x 10^y = 10^{x+y},$$

yielding that $\log ab$ is $x + y = \log a + \log b$. Similarly, the subtraction property follows from the Second Law, while interpreting the Third Law gives the additional property that $\log(x^y) = y \log x$ so that, for example,

$$\log\sqrt{10} = \log(10^{1/2}) = \frac{1}{2}\log 10 = \frac{1}{2}.$$

All that was required was a once-and-for-all calculation of a logarithm table for numbers between 1 and 10 and then logs of any number could be effectively looked up, as numbers outside the range 1–10 are easily dealt with through redress to the log laws. For example:

$$\log 84 = \log(10 \times 8.4) = \log 10 + \log 8.4 = 1 + \log 8.4 = 1.9243.$$

You may recall these two parts of the log, $\log 10$ and $\log 8.4$, being known as the *mantissa* and *characteristic* respectively.

Logarithms were all-important practical tools not very long ago and the slide rule was their physical manifestation. These devices were logarithmically scaled rules, machined with a fine degree of accuracy, for adding and subtracting logs. The better slide rules were beautiful pieces of engineering. If you still have one, you might be wise to cherish what could turn into a highly collectable item.

The techniques involved in logs are no longer taught at all as their primary purpose was practical and so they have become obsolete with the arrival of calculators which can do the job more quickly and accurately. However, a real loss accompanied the demise of the table book. Repeated poring over pages of log and trigonometric function tables did engender a familiarity with the behaviour of the functions themselves. What is more, the methods employed the use of scaling and interpolation (the estimation of intermediate values not explicitly tabulated), and so users needed to keep their mathematical wits about them to an extent that modern students, relying on a calculator, do not; once a problem has been reduced to a calculator application, the student becomes relatively passive, learns less, and is likely to accept anything the calculator displays quite uncritically.

I must add that the logarithm function remains important in science. Many natural scales are logarithmic—the pH scale of acidity, the Richter Scale of earthquake measurement, and the decibel scale of sound, to name three. In addition, the *natural* logarithm arises irresistibly in calculus—logarithms to the base $e = 2.7182...$, the number e being irrational and arising in problems concerning compound interest. Consequently science students still need to be thoroughly familiar with logarithms and they suffer through missing the traditional practical training that the log table afforded.

The invention of logarithms was a major boost to science around the turn of the seventeenth century and credit is due mainly to the Scot, John Napier. However, its development was not as straightforward as we might expect. Napier's original logs were much closer to the the so-called natural logs referred to above. Moreover, parallel techniques were in use by the astronomers Brahe and Kepler in Denmark about this time, whereby they performed exceedingly difficult calculations on the orbit of Mars using a technique involving trigonometric identities which also served to express products as sums. The importance of these identities came to be appreciated in Europe during the sixteenth century, the rules themselves having been discovered in the Middle East as far back as the eleventh century.

Since we are on the theme of irrationality, it is fair to mention that one of the difficulties with logarithms is that the log of a rational number, unless it is a power of 10, is irrational. For example, it is easy to see this for $\log 3$: once again we shall employ proof by contradiction. Suppose that $\log 3$ is equal to the fraction $\frac{a}{b}$: that would mean that $3 = 10^{a/b}$ and raising both sides to the power b would give $3^b = 10^a$. But this is impossible, as the left-hand side of this equality is odd while the right-hand side is even.

Irrationality is the norm

Although there may well be infinitely many true statements in the world, we all know that truth is a lot harder to come by than falsehood. In the same way, irrationality is any number of times more common than rationality when it comes to arbitrary numbers. This should not be taken as a disparaging remark concerning irrational numbers, but merely as a way of conveying the idea that, although there are infinitely many rational numbers, rationality of a number can truly be regarded as exceptional.

If we do think of numbers as decimal expansions, it becomes plain that irrationals, which have non-recurring expansions, must be much more common than the recurring expansions of the rationals. A naive argument is to imagine a random decimal somehow being generated (by picking digits out of a hat, say). The chances of the expansion falling into a recurring block pattern that repeats itself not just many times but *for ever* must surely be zero. This is actually a valid intuition, but one that would involve some work to make precise. The difficulty lies in that the argument confounds aspects of finiteness and infiniteness in that we allow ourselves to speak about the result of an infinite process as if we had actually carried it out.

A rebuttal might rest on the observation that both the sets of rationals and irrationals are clearly infinite so that it cannot make sense to say that one set could be larger than the other. This conclusion rests on the premiss that all infinite sets are essentially the same—an idea that does not stand up to serious scrutiny.

The peculiar nature of the infinite was first spelt out by Galileo. An infinite set can itself be divided into two parts, both of which are infinite and can be put into one-to-one correspondence with the set itself. For example, we need look no further than the set N of natural numbers, $\{1, 2, \ldots\}$. This can be partitioned into E and O, the sets of even and odd numbers respectively. In one sense, although all these sets are infinite, N is larger than E, as E is contained inside N. Galileo pointed out that what makes infinite collections different from finite ones is that one can remove infinite sets from them, like E from N, and what remains (O in this case) can still be infinite in the same way. Finite sets never behave like this—if we take something away from a finite collection, what is left behind is certainly smaller. This is the key difference in nature between the infinite and the finite. In practice it can make infinite sets easier to work with than finite ones once you become used to this aspect of their make-up.

There is another fundamental way in which infinite sets can differ from one another which is far less obvious, and seems not to have been appreciated until the end of the last century. Some infinite sets can be written in a list, and some cannot.

The set of natural numbers, called N, is the usual collection of counting numbers: $\{1, 2, 3, \ldots\}$. This set embodies the very idea of an infinite list. However, some other infinite sets can be put into one-to-one correspondence with the natural numbers and so be listed as well. For instance, take the set Z of all *integers*, that is the positive and negative whole numbers together with 0:

$$Z = \{\ldots, -3, -2, -1, 0, 1, 2, 3, \ldots\}.$$

This set comes to us naturally as a kind of doubly infinite list. It can, however, be rearranged into a list with a starting point as follows:

$$Z = 0, 1, -1, 2, -2, 3, -3, 4, -4, \cdots \qquad (2)$$

In fact, we shall make use of the idea used here more than once: if we have two lists:

$$a_1, a_2, a_3, \ldots, \quad b_1, b_2, b_3, \ldots,$$

we can *interleave* them to form a single list which comprises all the members of the original two:

$$a_1, b_1, a_2, b_2, a_3, b_3, \ldots$$

This is really what we did to combine the positive and negative whole numbers into a single list. You may think there is nothing much going on here. Given any set, surely it is possible to consider it as a list in some way? But how about the set Q of all rational numbers? (Why is Q used for rationals? It is something to do with the word 'quotient'.) In fact, it can be done, but we need to be cleverer. We shall look at this trickier problem in a moment. First I would like to remove a source of possible confusion.

The reader might well make the objection that I raised before, in that the above interleaving argument involved our talking loosely about an infinite process as if we had actually carried it out in its entirety. This viewpoint is not necessary in order to show, for instance, that the set of integers can be listed, provided we clarify what we mean. By forming an infinite list L, what I shall mean here is that for every number n there is a rule that specifies the nth term in L. When I claim that L is a list of all integers, what I mean is that, given any particular integer k, it is possible to find the place in the list L where k comes up. In other words, although we would have to wait for ever for *all* the integers to come up, we only have to wait for a finite number of steps for any named integer to appear. It is true that I never gave the rule for specifying the nth member of the list (2) above explicitly, but rather relied on readers recognizing the simple pattern used to construct it. There is nothing wrong with this provided that we can always continue to write down more of the list in an unambiguous way. We do not have to cheat, however. The positive number n occupies the $2n$th place in the list—for example, 3 is sixth in the list—and the negative number $-n$ is in the $(2n + 1)$th position—for example -3 is in the seventh position, and 0 is placed first. We see that we therefore know the exact position of each integer in our list—they are all present and accounted for.

Now let us consider the problem of listing all the rationals between 0 and 1. This looks a tough job as the rationals are *dense*, meaning that between any two of them there is another; for example, their average lies exactly halfway between the two. This, however, causes no real trouble as long as we do not insist that we write our numbers in increasing or decreasing order—we simply list all the rationals with 1 as a denominator first (that is, $0 = \frac{0}{1}$ and $1 = \frac{1}{1}$), then all those with denominator 2, all with denominator 3, and so on:

$$0, 1, \frac{1}{2}, \frac{1}{3}, \frac{2}{3}, \frac{1}{4}, \frac{3}{4}, \frac{1}{5}, \frac{2}{5}, \frac{3}{5}, \frac{4}{5}, \cdots$$

The key observation is that there are only ever finitely many rationals between 0 and 1 with a given denominator (if n is the denominator, there are no more than n of them), so building a list in this way will eventually capture every rational between 0 and 1. None will escape.

Now comes another trick. If we take all the members of this list apart from 0 and 1 and invert each of them, we obtain a list of all the rational numbers that are greater than one:

$$\frac{2}{1}, \frac{3}{1}, \frac{3}{2}, \frac{4}{1}, \frac{4}{3}, \frac{5}{1}, \frac{5}{2}, \frac{5}{3}, \frac{5}{4}, \cdots$$

This requires a little thought. Since all the fractions in the first list lie between 0 and 1, their reciprocals are all greater than 1. Moreover, if $\frac{m}{n}$ is a rational greater than 1, then $\frac{n}{m}$ is a rational less than 1, so that it occurs somewhere in our first list—therefore its reciprocal, $\frac{m}{n}$, will occur in the corresponding place in the second list. For example, $\frac{5}{4}$ lies in the ninth position in the inverted list as $\frac{4}{5}$ is ninth in the list of fractions (beginning at $\frac{1}{2}$). Again, no rational misses out. This may seem too good to be true, as there looks to be far more rational numbers greater than 1 than between 0 and 1. However, as I said, infinite sets can be strange.

Now we have two sets that can be listed: the rationals between 0 and 1, and the rationals greater than 1. Employing the interleaving argument we used previously to show that all the integers can be listed, we can combine these two sets into a single list, thus showing that all the rationals from 0 upwards can be listed.

Finally, by the same method, we can also list all the negative rational numbers and, interleaving once again, we can combine this list with that for the non-negative rationals to produce a single list which contains every rational number. We can actually write down the first couple of dozen numbers in our list: we write the rationals that are whole numbers without their denominator of 1 in order to save on clutter. To keep the display more symmetrical, we shall arrange things a little differently. Begin with 0 and let the first list L_1 be that for the rationals between 0 and 1; let L_2 be that formed by inverting the members of L_1; let L_3 be all the negatives of the members of L_1; and let L_4 consist of all the negatives of L_2. The interleaving process then gives our grand list for Q as follows:

$$0, 1, -1, 2, -2, \frac{1}{2}, -\frac{1}{2}, 3, -3, \frac{1}{3}, -\frac{1}{3}, \frac{3}{2}, -\frac{3}{2}, \frac{2}{3}, -\frac{2}{3}, 4, -4, \frac{1}{4}, -\frac{1}{4}, \frac{4}{3},$$

$$-\frac{4}{3}, \frac{3}{4}, -\frac{3}{4}, 5, -5, \frac{1}{5}, -\frac{1}{5}, \frac{5}{2}, -\frac{5}{2}, \frac{2}{5}, -\frac{2}{5}, \frac{5}{3}, -\frac{5}{3}, \frac{3}{5}, -\frac{3}{5}, \frac{5}{4}, -\frac{5}{4}, \frac{4}{5},$$

$$-\frac{4}{5}, \dots$$

Readers should not have too much trouble extending this list to a further dozen terms or more.

To infinity and beyond!

Buzz Light Year of the children's film *Toy Story* exhorts us to travel to infinity and beyond, a thing very dear to the heart of mathematicians who have made it their business to do just that for over a century and now have a pretty good idea of what to expect upon arrival.

There are many large sets of numbers which can be listed in the way described in the previous section. One such set, which contains Q, is the collection of all *algebraic* numbers. These are the numbers that are solutions to polynomial equations with integer coefficients (that is to say, equations such as $6x^3 + 5x^2 + 8x + 1 = 0$, where the numbers multiplying the powers of x are integers). Every rational $\frac{a}{b}$ is the solution to the simple equation $bx - a = 0$, and so is algebraic. The same applies to $\sqrt{2}$, which is a solution of the quadratic equation $x^2 - 2 = 0$, and, for that matter, to $2^{\frac{1}{3}}$, the cube root of 2, as it is a solution of $x^3 - 2 = 0$. A number that is not algebraic is somewhat mysteriously called *transcendental*. As we shall soon see, transcendental numbers are not at all rare, although it can be extraordinarily difficult to verify that a particular number is transcendental. Our irrational number b introduced earlier is transcendental (although this is far from obvious), and so is π. That π is not algebraic was proved last century by Lindemann, and a consequence of this is the impossibility of squaring the circle; that is, given a circle, it is not possible to construct a square, using a straight edge and compasses, with the same area as the given circle. The difficulty lies in the fact that all constructible numbers are algebraic. The squaring of the circle is in effect a challenge to construct $\sqrt{\pi}$. If you could construct $\sqrt{\pi}$ then you could construct the transcendental number π, but it is impossible to construct a transcendental number.

Not even all algebraic numbers are constructible. In particular, $2^{\frac{1}{3}}$ is an algebraic non-constructible number and this settles another classical question: given a cube, can one construct another cube of exactly twice the volume of the original? This is the famous Delian Problem, the task set by the god in order to banish the plague from Athens.

The final of this trio of classical problems is the task of constructing the trisection of an arbitrary angle. It turns out that, although it is simple enough to construct a 60° angle, it is not possible to construct a 20° one. And so all three of these questions were answered in the negative, although around 2,200 years after they were originally posed.

Many people feel affronted by the word 'impossible' and refuse to believe any scientific pronouncement containing it. The above claims can be made in a less provocative manner as follows. It turns out that constructible numbers have certain special properties that not all numbers enjoy, and it can be verified in particular that $2^{\frac{1}{3}}$ lacks one of these properties. This mild statement is effectively the same as the bolder assertion that it is impossible to duplicate the cube.

Returning to our current investigations, we have listed the collection of all rational numbers, that is to say all numbers with recurring decimal expansions. We shall now show that it is not possible to make a similar list of all the real numbers—all decimal expansions, if you like, of the numbers—between 0 and 1. How do we know that there is not some cunning way of doing it that we simply have not yet thought of? We know because Georg Cantor, in the late nineteenth century, invented his so-called *Diagonal Argument* proving this to be impossible. All it consists of is the observation, explained more carefully in a moment, that, given any infinite list L of decimals (between 0 and 1, let us say) it is possible to use the list itself to construct another decimal between 0 and 1 that does not lie in the original list L. This sounds fairly innocuous, but it follows immediately from this that there is no list L which contains every real number between 0 and 1.

The argument itself runs like this. Suppose you have your list L. All we need do is write down a number a which differs from the first number in L in the first decimal place, differs from the second in the second decimal place, and so on, differing from the nth in the nth decimal place. The number that you so construct differs from every other one in the list. If you imagine all the decimal displays in L listed one under another, we build the number a by looking down the diagonal of the display from top left to bottom right and ensuring that a differs from the nth row of the array at the entry that lies in the nth column.

Sets that cannot be listed are called *uncountable*; sets that can be listed are *countable* (even though they may be infinite, such as the set of rationals). Clearly, if a set A is countable, so is any set B contained in it, since in order to list B we need only take the list for A and read through, building a list for B as we go along. It follows that, if S is an uncountable set, then so is any set T that contains S (for if T were countable, then S would be by the previous argument). Therefore, since the set of real numbers between 0 and 1 has proved to be uncountable, it follows that the set R of all real numbers is uncountable, even though the set Q of all rational numbers is countable. We have thus discovered a new qualitative way in which the set of all decimals is a greater collection than the set of rationals.

More can be said. It follows from the fact that the set of all algebraic numbers A is countable (something that we have not shown here, but is only a little more difficult than showing that the rationals are countable) that the set T of all transcendental numbers is uncountable. (If T were countable, we could use the interleaving argument to show that the union of A and T was countable, but this is a contradiction as that union is the set of all real numbers, which we now know is uncountable.) This is a very impressive result: it establishes that T is an uncountable (in particular, an infinite) set without actually exhibiting any members of it. In other words, we can know that there are uncountably many transcendental numbers without necessarily knowing the identity of a single one of them!

3

Some Geometry

In this chapter I aim to provide demonstrations of a few of the more famous results of Euclidean geometry, including the theorem of Pythagoras and some of the Circle Theorems. The proofs remain as surprising and delightful today as they were thousands of years ago, and we can be sure that our own distant progeny will also be charmed by their elegance.

I will begin with Pythagoras. This theorem links geometry and algebra like no other single fact. It gives an algebraic hold on the physical notion of distance and so its presence is felt throughout mathematics and physics—the Theory of Special Relativity, for instance, rests upon it.

The importance of squares on sides of triangles

The theorem of Pythagoras says that the square of the hypotenuse c of a right-angled triangle is equal to the sum of the squares of the other two sides, a and b (see Figure 1). This can be seen at once by comparing the two squares pictured in Figure 2. Each picture is that of a square of side length $a + b$, and so they represent equal areas. Each picture includes four copies of the given right triangle, so, if we remove these, the remaining shaded regions of the pictures have equal areas also. Evidently, the first shaded area is $a^2 + b^2$, while the shaded area on the right equals c^2. This completes the proof.

It truly is easy. I can think of no reason why everyone is not shown this demonstration in school. Indeed, if there is a fault in this proof, it is that it is over almost before you know it. A sceptical person might ask: exactly where did the fact that the triangle had a right angle come into it at all? Answering this question reveals that the proof does make at least one hidden assumption, which I shall explain now.

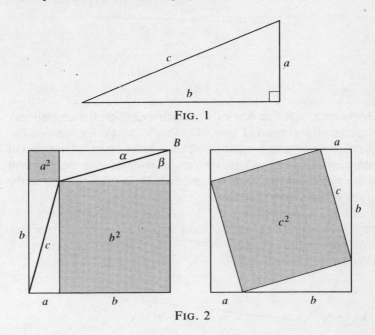

FIG. 1

FIG. 2

We do need to know that the three angles in the right-angled triangle add to that of a straight angle. (This is true of any triangle of course, as we shall see shortly.) This justifies the claim that the figure on the left and the shaded figure on the right are indeed squares. The angle at B, for instance, must be a right angle as it is the sum of the two acute angles, α and β, of the right triangle which must sum to $180 - 90 = 90°$; so let us prove this more fundamental fact about triangles.

To do this, we need some still more basic properties of angles.

Property 1. Where two lines meet, opposite angles are equal

(Figure 3). That is to say, the two angles *A* and *B* are equal: this is because both $A + C$ and $B + C$ equal a straight angle.

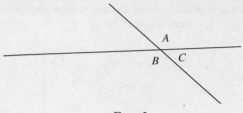

FIG. 3

Property 2. Where a line cuts two parallel lines, the corresponding angles are equal (Figure 4). This is an axiom, one of our fundamental starting rules for which we give no proof in terms of other assumptions. (Any system of mathematics begins with some assertions that are not proved. Pure mathematics is the study of their consequences.)

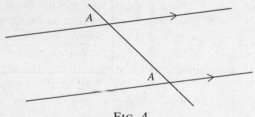

FIG. 4

Now let *ABC* denote an arbitrary triangle, where we have called the associated angles α, β, and γ (the Greek letters alpha, beta, and gamma—I fear brows will furrow at the introduction of such symbols, but take a breath and don't be intimidated by them). We want to prove that $\alpha + \beta + \gamma = 180°$, the value of two right angles. In Figure 5, let *L* be the line through *B* parallel to *AC*, and extend the lines *AB* and *BC* as shown. We are entitled to mark the three angles above *L* as indicated, for the β is justified by Property 1, while Property 2 explains the other two: compare the two angles marked α and the two angles marked γ and remember that L is parallel to AC. It remains only to observe that the three angles in question comprise the straight angle at the point B on L to gain the required conclusion that $\alpha + \beta + \gamma = 180°$. This is what we set out to prove.

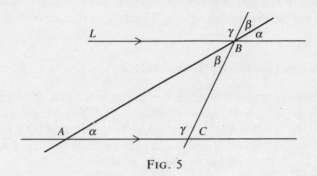

FIG. 5

We have therefore set Pythagoras on a firm foundation. One elementary algebraic fact comes out of this geometry. The area of our right-angled triangle is $\frac{1}{2}ab$: you do not even need the famous $\frac{1}{2}$base × height formula to see this, as two copies of the triangle clearly form a rectangle of area ab. Hence the four triangles in each of our two large squares have a combined area of $2ab$. It follows that the square on the right of our original picture has area $(a+b)^2$, and that is also equal to $c^2 + 2ab$. Now use Pythagoras to replace c^2 by $a^2 + b^2$ to obtain:

$$(a+b)^2 = a^2 + b^2 + 2ab. \tag{1}$$

This is a simple algebraic fact of which we shall have more to say in Chapter 5. If you are prepared to use it as a starting point, then you can derive Pythagoras from the picture of the square on the right in Figure 2 alone, for writing its area as the sum of its parts gives:

$$(a+b)^2 = c^2 + 2ab,$$

and using our identity (1) we obtain:

$$a^2 + b^2 + 2ab = c^2 + 2ab.$$

We now need only remove the unwanted $2ab$ from both sides of the equation to recover Pythagoras.

This proof, which comprises a mixture of geometry and algebra, is perhaps a little less pretty than our original, but it does have the advantage of being easy to remember: you only need to recall the picture on the right of Figure 2 to reproduce it.

Pythagoras reveals truth about circles

The force of a result is often not apparent at first sight and it would be not unreasonable, despite my general assurances earlier, to say that Pythagoras's Theorem seems to be a rather dull fact about a very special form of triangle. Why should we be interested in drawing squares on the sides of a triangle in the first place?

Experience has shown that the Pythagorean relation arises constantly in mathematics and physics and neither subject would ever have got anywhere without it. As an example, an unexpected consequence of the theorem was discovered by Herodotus in the fifth century BC. By this stage of its history, mathematics was already considering some quite sophisticated questions. In particular, finding exact areas of figures with curved boundaries had proved elusive. Much of this was bound up with the mysterious nature of the number π, which was not to be resolved for thousands of years. It was hard therefore to resist the pessimistic conclusion that it was impossible to find the *exact* area of any figure whose boundary was curved or partly curved. Herodotus showed that this was not the case by devising a series of clever examples of *lunes*, crescent moon shapes, whose area could be found exactly. The first of these comes from pondering a little more about what Pythagoras really says.

The square mounted on the hypotenuse of a right triangle has area equal to the sum of the areas of the squares on the shorter sides. However, the same is true if we mount semicircles. Why so? The area of a semicircle of radius r is $\frac{\pi}{2}r^2$ and so the area of the semicircle on the side a is $\frac{\pi}{2}\left(\frac{a}{2}\right)^2 = \frac{\pi}{8}a^2$, with similar expressions for the semicircles on sides b and c. However, since $a^2 + b^2 = c^2$

we obtain $\frac{\pi}{8}a^2 + \frac{\pi}{8}b^2 = \frac{\pi}{8}c^2$, showing us that the area of the semicircles on the shorter sides sum to the area of that on the hypotenuse. And there is nothing special about semicircles either: we could replace the squares with any similar figures whose areas were proportional to the square of the side length.

That said, consider Figure 6. It involves a unit square $ABCD$ with a semicircle ABC mounted on the diagonal AC. The mid-point of AC is M. We have also drawn a quarter circle with radius DA from A to C. We shall now calculate the area of the shaded lune.

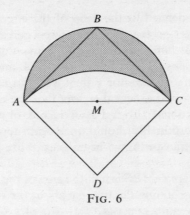

Fig. 6

The area of the lune equals the area of the triangle ABC for the following reason. We obtain the lune by taking the triangle, subtracting the segment of the larger quarter circle with chord AC, and then adding the two smaller outside segments. Now the larger segment is similar to each of the two smaller segments (that is to say, it has the same shape, only larger) because the triangles ADC and AMB are similar, each being an isosceles right triangle. It follows from Pythagoras that the area of the segment on the hypotenuse of ABC equals the sum of the areas of the segments on the shorter sides so that the net effect of the additions and subtraction is zero. We conclude that the area of the lune is the area of the triangle ABC which is $\frac{1}{2} \times 1 \times 1 = \frac{1}{2}$.

What Herodotus had shown was that is was possible, at least sometimes, to find the area of a figure bounded entirely by arcs of circles. This example is even a constructible figure—it is possible to draw the segments using only straight edge and compasses. This must have engendered hope that it might be possible to square the circle, as it was realized that, if the area of *any* lune could be found, then the area of a circle, and thus π, could also be determined. There are limitations with this method, however, and it has been suggested that Herodotus himself appreciated this. He none the less devised other so-called lunar quadratures.

Triangles and areas

The number π is defined as the ratio of the circumference of a circle to its diameter, so that the circumference of a circle is $2\pi r$, r being its radius. Certainly, doubling all the linear dimensions of a plane figure increases its area by a factor of 4, and in general if we enlarge a figure by a factor c then we enlarge its area by a factor of c^2. We would expect therefore that the area of a circle would be proportional to r^2, but it is not obvious why the constant of proportionality should once again turn out to be π. To see why this transpires, we begin once more by looking at triangles.

The area of a triangle ABC is half the area of the parallelogram $ADBC$ formed by rotating the triangle about the side AB, as the resulting parallelogram consists of two copies of the original triangle (Figure 7). The area of the parallelogram is bh, where h is

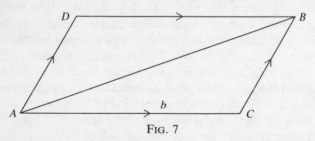

Fig. 7

the height of the triangle. This can be seen because a triangle may be cut from one end of the parallelogram and replaced at the other to form a bh rectangle as shown in Figure 8. It follows that the area of a triangle ABC is $\frac{1}{2}bh$. What is surprising is that this

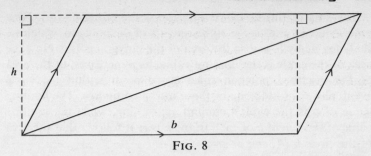

FIG. 8

half-base times height formula also holds for a circle if we regard
the base as the circumference and the height as the distance from
the circumference to the centre:

$$\frac{1}{2}bh = \frac{1}{2}(2\pi r)r = \pi r^2.$$

This suggests that we should try to establish the area of a circle
through some kind of triangulation of the figure.

Take n equally spaced points around the circumference of
the circle, thus forming a regular n-sided polygon within our
circle. (By 'regular' polygon we mean that all sides and all angles
are equal.) By joining each point to the centre of the circle, we
break the polygon into n triangles of equal height h and base B
(see Figure 9).

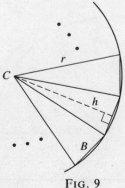

FIG. 9

The area of this inscribed polygon is therefore $n(\frac{1}{2}Bh)$. Now nB is evidently the length of the outside of the polygon, which we shall denote by b, and so the area of the polygon is $\frac{1}{2}bh$. Now the area of the circle is the limiting value, as n increases, of the area of the inscribed polygon since every point within the circle eventually sits inside one of these inscribed figures. The limiting value of b is the circle's circumference, $2\pi r$, while the limiting value of the height h of each triangle is r. It follows that the area of the circle is $\frac{1}{2}(2\pi r)r = \pi r^2$.

This is a convenient juncture at which to mention the measurement of angles. The practical approach is to divide the circumference of a circle into 360 units of turn known as 'degrees'. This is somewhat arbitrary but the number 360 has many factors, so that most simple fractions of a circle correspond to a whole number of degrees. Moreover, a rotation of one degree is about the smallest that is noticeable to the unaided eye, which makes it a useful unit of measurement. However, if you are more interested in properties of geometrical objects than in their measurement, another unit is more relevant. The length of the circumference of a circle of radius one unit is 2π, so it is mathematically simple to put one unit of turning equal to one unit of travel around the circumference. This unit of angle measurement is known as the *radian*, and so there are 2π radians in a circle (Figure 10). The measure of a straight angle is then π radians, while a right angle equals $\frac{\pi}{2}$. One radian turns out to be a little over $57°$.

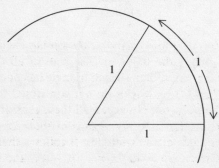

FIG. 10

It makes little difference to what we are about to do, but a single symbol, π, demands less space than writing 180°, so for that reason we shall use radian measure in this chapter unless we explicitly say otherwise.

This idea of triangulation, of partitioning a geometrical object into triangles in some particular way, may seem quite naive and simple-minded, but is very fruitful in geometry and topology, the branch of mathematics that deals with the general properties of shapes and space. It can be rather striking to see how many very difficult problems in mathematics can successfully be tackled through patient build-up from special cases to general ones.

In passing I should mention that, once we know that the angles of a triangle sum to π radians (180°), it is simple enough to calculate the sum of the angles of any polygon. For example, take any regular n-sided plane figure P (in Figure 11 we take $n = 6$).

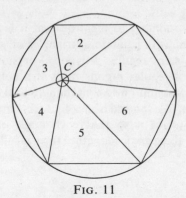

FIG. 11

We can partition P into n triangles by joining each vertex to a single point C inside the polygon. The sum of all the angles of the triangles is then $n\pi$ radians. Each triangle has one of its angles at the vertex C and these angles do not contribute to the sum of the angles of the polygon. However, all these central angles form one complete turn; that is to say, they contribute 2π (360°) to the sum of all the angles of the triangles. It follows that the sum of the angles of P must equal $n\pi - 2\pi = (n - 2)\pi$, and in particular the sum of the angles in any quadrilateral (a four-sided polygon) is 2π radians or 360°. Of course, if we take $n = 3$ we recover the sum of the angles of a triangle: $(3 - 2)\pi = \pi$.

Circle Theorems

Hexagons

A *circle* of radius r and centre C is the set of points in the plane whose distance from C is equal to r. Circles are the most symmetrical objects imaginable in the plane and so it is to be expected that the circle is brimming with special properties.

One such property relates to hexagons. As has long been known to honey bees and makers of patchwork quilts, it is possible to *tessellate* the plane with hexagons—that is, the plane may be covered by identical hexagons in such a way that no two overlap except at their boundaries—and this property is exploited in their work.

I digress for a moment to mention that it is also possible to tessellate the plane with equilateral triangles or with squares, but with no other type of regular polygon. For suppose that some number, k say, of regular n-gons meet at a common point the way they do in a tessellation so that the sum of their angles is that of a full circle, 2π. We showed above that the sum of the angles of the polygon is $(n - 2)\pi$ so that each angle equals $\frac{(n-2)}{n}\pi$. It follows that k of these angles together form a full circle; that is to say:

$$\frac{k(n - 2)}{n}\pi = 2\pi,$$

or, what is the same, $k = \frac{2n}{n-2}$. However, the number k is an integer and the expression on the right is integral only for $n = 3$, 4, or 6: for $n = 5$ we get $\frac{10}{3}$, while for any number larger than 6 the expression lies strictly between 2 and 3. Hence there are no other tessellations by regular polygons except the three obvious ones. There are many other interesting tessellations not of this type, however. Copies of any triangle can cover the plane (form parallelograms using the triangle as we did above and then it is easy to see how to do it), octagons and squares together can form a complete covering, while hexagons and pentagons together can form a spherical shape such as a soccer ball. Some years ago Roger Penrose of Oxford University showed that it was possible to cover the plane with copies of two simple but rather irregular shapes in such a way that the pattern was never repeated—in other words, the tessellation looks different from wherever you place yourself in the plane.

Returning to our hexagons, quilters are often advised to construct the basic hexagon for their backing sheet by drawing a circle and stepping the radius into the circumference six times to give them the vertices of the hexagon. I did once overhear someone (not a quilter) explain this somewhat apologetically, saying that this method was not exact but was adequate in practice, and the confused instructor went on to say, in so many words, that the inaccuracy is due to 2π not being exactly equal to 6. Not at all! It *is* exact (even though 2π is indeed greater than 6) and one can easily see why, as follows.

Open your compasses out to the radius r of the circle, put the compass point on A, any point on the circumference you wish, and mark a point B where the tip of the compasses cuts the circle. We call the centre of the circle C, so we have the situation shown in Figure 12. The key observation is that, since A, B, and C are all the same distance r from one another they form an equilateral triangle. In particular, the angle ACB is 60°, which is exactly one-sixth of the full circle's angle measure. It follows that stepping out the radius five more times will return you *exactly* to your starting point A and the six points on the circle will form a regular hexagon.

This is perhaps the simplest of a number of pretty symmetries

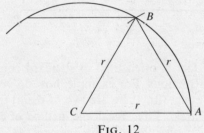

FIG. 12

of the circle known collectively as Circle Theorems. Although some of these theorems are quite surprising, their proofs exploit the characteristic property of the circle over and over, that property being that points on the circumference are always the same distance r from the centre, together with the fact that the angles of any triangle sum to 180°.

Angles in semicircles

The next example of a Circle Theorem is that the angle in a semicircle is a right angle. This means that, if A and B are the endpoints of a diameter of a circle and D is any other point on the circle, then the angle ADB is a right angle. In other words, as the point D travels around the circumference of the circle, although the distances AD and BD change, this angle never does—the two lines always meet at a right angle (see Figure 13). This can easily be seen through joining the point D to the centre C, thus partitioning the large triangle into two smaller isosceles triangles, the three marked sides all being of common length r. Being isosceles, each has a pair of equal angles, marked α and β respectively in Figure 13. We see the angles of the larger triangle sum to $\alpha + \alpha + \beta + \beta = 180°$ and so $\alpha + \beta$, which equals the measure of the angle at D, is 90°.

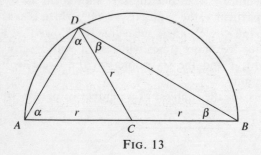

FIG. 13

This particular fact is often used in standard compasses and straight-edge constructions. For example, how do you construct the tangents to a circle (for there are two of them) from a given point outside the circle? First, recall the method of construction of the perpendicular bisector of a line segment AC as shown in

Figure 1 of the previous chapter. Find the centre of the circle by taking the intersection of the perpendicular bisectors of any two chords of the circle. Join the centre C to the given point P and find the centre O of CP (again, through bisection) (Figure 14). Draw the circle with centre O and radius OP; it will cut the original circle at two points T and S and the lines PT and PS are the required tangents.

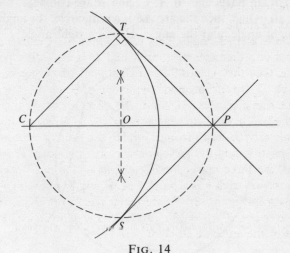

Fig. 14

Why? The characteristic property of a tangent to a circle is that it forms a right angle with the radius at the contact point. The angle CTP is a right angle as T (and similarly S) lies on the circumference of a circle with CP as diameter.

That the angle in a semicircle is a right angle is both a striking and useful fact, but it is only a special case of the next theorem.

Angles at the centre of circles

The angle at the centre of a circle is twice that at the circumference standing on the same arc; that is, $\angle ACB = 2\angle ADB$.

This theorem may need some explanation, and the accompanying diagrams should go some way towards doing this. Let

AB be any arc of a circle as shown in Figure 15. By the *angle at the centre of the circle* standing on this arc, we mean the angle *ACB*. Now let *D* be any point at all on the circumference outside the sector of the circle *ACB*. The angle *ADB* is what is meant by the *angle at the circumference* standing on the arc *AB*. The theorem is then saying that, as *D* moves around the circle from *A* to *B*, this angle never varies, being always equal to one-half of the angle *ACB*. In particular, if *A*, *C*, and *B* are collinear, meaning they are all in line, then the arc *AB* is a semicircle, the angle *ACB* is a straight angle, and the angle *ADB* is a right angle.

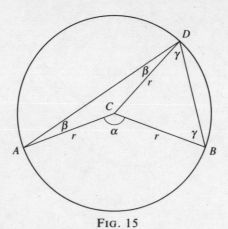

FIG. 15

To see why the theorem is true, look at the picture and note the way in which we have, as in the semicircle theorem above, marked the radii of the circle and labelled pairs of equal angles. Our task is to check that $\alpha = 2(\beta + \gamma)$. Since angles within any triangle sum to π, we infer that the unmarked angles at *C* have values of $(\pi - 2\beta)$ and $(\pi - 2\gamma)$ respectively. Since the three angles at the centre of the circle add to 2π, we have:

$$(\pi - 2\beta) + (\pi - 2\gamma) + \alpha = 2\pi$$
$$\Rightarrow 2\pi + \alpha - 2\beta - 2\gamma = 2\pi,$$

whereupon, subtracting the unwanted 2π from both sides yields:

$$\alpha = 2\beta + 2\gamma = 2(\beta + \gamma).$$

This is the proof but not the whole proof, as Figure 15 does not represent all cases. As the point D moves, let us say, clockwise around the circle, there will come a point where A, C, and D lie in a line. The above argument copes with this situation—the angle β is equal to 0 but nothing else changes. However, as D continues to travel around the circle the centre C appears outside of the triangle ABD, as shown in Figure 16. This represents a genuinely

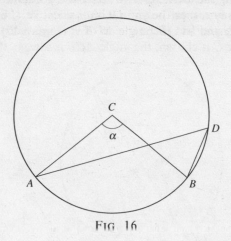

FIG. 16

different case, and a different (though similar) argument, which I omit, shows that the angle α remains equal to twice the angle ADB.

There is one remaining facet of the situation. Return to Figures 15 and 16. As D moves around the circumference from A to B, the angle ABD is always $\frac{\alpha}{2}$. However, as D passes through B there is a discontinuity—a sudden jump if you like. The angle ADB is still half the angle at the centre of the circle, but now it is the reflex angle ACB that applies. For example, measuring in degrees, let us return to Figure 15 and suppose that the angle $ACB = 140°$. Then the angle ADB equals 70° until D actually passes B and enters the lower sector of the circle, when it suddenly becomes

$$\angle ADB = \frac{1}{2}(360 - 140)° = \frac{1}{2}220° = 110°.$$

It then stays at this value until D passes through A, when it returns to 70°.

The fact that the angle ADB is the same for any point D is often expressed by saying that two angles 'standing on the same arc', the arc AB in this case, are equal. This has a consequence for regular polygons, although the connection may not be immediately apparent. We shall have reason to recall it later when studying the Golden Ratio so we draw attention to it here.

Let P be any regular polygon with n sides, let C be one of the corners of P, and let AB be a side of P (Figure 17). No matter which corner C is chosen, the angle ACB is always the same and

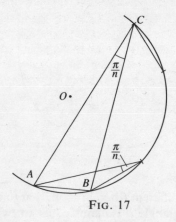

FIG. 17

is equal to $\frac{\pi}{n}$. To see why this is so, take a circle and imagine forming a regular n-sided polygon by taking n equally placed points around the circumference of the circle. Taking a side AB and a corner C as above, we see that angle ACB is the angle standing on the arc AB of the circle and so equals half of the angle at the centre. Since the points of P are equally spaced around the circle, the angle at the centre is $\frac{2\pi}{n}$ and so the angle ACB is $\frac{\pi}{n}$ as claimed.

Another result that is now easily demonstrated is that the opposite angles of a so-called *concyclic* quadrilateral add to 180°. A concyclic quadrilateral Q is one that can be inscribed in a circle. Not all quadrilaterals have this property. It is true that any *three* points that are not *collinear*, that is do not lie on the one line, lie on a unique circle, and it is easy to construct that circle as follows.

The centre *O* of any circle through the points *A, B*, and *C* is equidistant from all three points. Now the perpendicular bisector of the line segment *AB* consists of all those points that are equally distant from *A* and *B*. Hence we see that *O* must lie on this bisector and, by the same reasoning, *O* must lie on the bisector of *BC* as well. Therefore *O* is the common point of the perpendicular bisectors of *AB* and *BC* (Figure 18). Equally, of course, *O* must lie on the perpendicular bisector of *AC*, so the three bisectors are *concurrent*; that is, they are lines which meet at a common point. Regarding *ABC* as defining an arbitrary triangle, we see that we can interpret this construction as saying that the perpendicular bisectors of the sides of any triangle are concurrent.

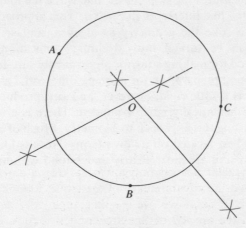

FIG. 18

Now, for any quadrilateral $Q = ABCD$, the sum of its angles is 360°. As we have just seen, there is a unique circle having *ABC* as an arc and, in general, there will be no reason why the fourth corner *D* should also lie on the circle. If, however, it does, we say that *Q* is concyclic and *Q* then enjoys the additional property mentioned above: each pair of opposite angles is *supplementary*, that is sums to 180°. Readers can easily convince themselves of this by drawing the appropriate picture and joining each corner to the centre of the circle, thus forming four isosceles triangles. Mark all the angles, equal angles being marked by the same symbol, and you will see that the sum of each pair of opposite angles is the same. Hence each pair must sum to a half of 360°.

Euclidean geometry is not much emphasized in schools at present. Geometry is viewed mainly through the perspective of so-called analytical geometry. This approach was invented by René Descartes and can claim much success. The basic idea is always to work within a *rectangular coordinate* framework—a pair of x and y axes if you like. Lines and curves are then treated through the equations that relate the coordinates of their points. For example, any straight line consists of all those points (x, y) in the plane that satisfy an equation of the form $y = mx + c$; the value of m measures the slope of the line while the number c tells you where the line intercepts the y-axis. This approach in effect allows us to code the geometry as algebra, and so geometric theorems can be turned into algebraic verifications. This is certainly good for making students algebraically sure-footed, and affords a sound preparation for learning differential and integral calculus. It is a little rigid, however, and can produce students who are mathematically narrow-minded. There is a real loss in exclusively using this approach to geometry. Much of elementary geometry is best tackled on its own terms—results like the ones we have just seen are more clearly understood without recourse to coordinates.

Vectors

A kind of halfway house between classical geometry and coordinate geometry is found in the use of *vectors*. The importance of this concept in mathematical physics is immense. Here I shall merely introduce the idea and give an example to show how they can be handy when it comes to explaining certain types of geometrical fact.

For our purposes, we shall regard a vector as an instruction to travel a certain distance in a certain direction. As such, a vector **v** can be thought of as an arrow where the length of the arrow is the distance to be travelled and the arrow points in the direction you are to go. We can add two vectors **u** and **v** together to form a new vector **w** = **u** + **v** (Figure 19). The arrow for **w** is obtained by first laying down **u**, then placing the tail of **v** on the tip of **u** and forming the arrow whose tail is that of **u** and whose tip is that of **v**. The vectors **u** and **v** can be added in the opposite order with the same net effect: in either case, the resulting vector sum **w** corresponds to the directed diagonal of the parallelogram as shown in Figure 19.

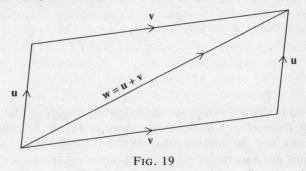

FIG. 19

We can also multiply vectors by numbers: 3**v**, for instance, would mean a vector with the same direction as **v** but three times as long; the vector −2**v** is twice the length of **v** and is taken to point in the opposite direction to **v** because of the presence of the minus sign (Figure 20). This gives us a little algebraic framework in which to work with our vectors. I do need to mention one more thing to make it complete. If we add **v** and −**v** we get back where we started. We represent this as the zero vector, **0**: it has a magnitude of 0 and, unlike all other vectors, has no direction associated with it. Since the vector −**v** makes sense, we can use it to define subtraction of vectors: **u** − **v** means **u** + ⁻**v**, rather like, in the case of ordinary numbers, 2 − 3 can be used to abbreviate 2 + (⁻3).

This is enough to show one of the basic argument styles used when working with vectors. We have some geometrical object and we look at two points A and B on the object. We then move around the object from A to B in two different ways, describing our paths as sums of two or more vectors. We then equate the two vector sums, as each is equal to vector **AB** from A to B. From this equation, less obvious equalities can be inferred.

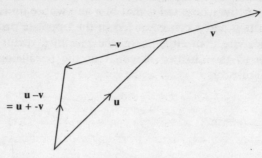

Fig. 20

As an example let us prove the following surprising fact. Take any quadrilateral $Q = ABCD$ as shown in Figure 21. Then we shall see that the quadrilateral $EFGH$ formed by joining the midpoints of the sides of Q is a *parallelogram*; that is to say, EF is equal and parallel to HG and FG is equal and parallel to EH.

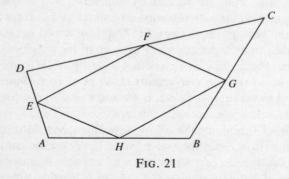

Fig. 21

To begin with, we prove the claim for one pair of sides, EF and HG say, and we note that this amounts to the assertion of equality of the two vectors **EF** and **HG**, so we look for a vector equation relating them. We can write one down immediately: travelling from A to C via E and F, and then from A to C via H and G, gives us two vector sums which are both equal to $\mathbf{AC}$:

$$\mathbf{AE} + \mathbf{EF} + \mathbf{FC} = \mathbf{AH} + \mathbf{HG} + \mathbf{GC}. \tag{2}$$

This looks promising, as (2) has $\mathbf{EF}$ on one side and $\mathbf{HG}$ on the other. There are, however, four other terms that we wish to get rid of. What is more, we have yet to use the fact that the points E, F, G, and H are the midpoints of the sides of Q, so there is a little more thinking to be done.

We do have another vector equation:

$$\mathbf{AD} + \mathbf{DC} = \mathbf{AB} + \mathbf{BC}.$$

Since E is the midpoint of AD, we can write $\mathbf{AD} = 2\mathbf{AE}$: we can similarly rewrite the other terms in the previous equation to get:

$$2\mathbf{AE} + 2\mathbf{FC} = 2\mathbf{AH} + 2\mathbf{GC}. \tag{3}$$

Halving the lengths of all vectors in (3) then shows us that:

$$\mathbf{AE} + \mathbf{FC} = \mathbf{AH} + \mathbf{GC}. \tag{4}$$

Finally, we re-examine equation (2). Since vectors can be added in any order, the equation can be written as

$$\mathbf{EF} + (\mathbf{AE} + \mathbf{FC}) = \mathbf{HG} + (\mathbf{AH} + \mathbf{GC}) \tag{5}$$

Equation (4) tells us that the vectors in brackets in (5) are equal: if we subtract them from both sides of (5) we get what we want:

$$\mathbf{EF} = \mathbf{HG}.$$

In the same way you can verify that $\mathbf{FG} = \mathbf{EH}$, thus showing that the midpoints of the sides of any quadrilateral do form a parallelogram.

Another example along these lines is the fact that the diagonals of a parallelogram meet at their midpoints. You may convince yourself of this by using a similar approach: begin at one corner and 'travel' to the midpoint of each diagonal, coding each journey as a vector sum. You then need to verify that these two vector sums are in fact equal so that the two diagonal midpoints do indeed coincide.

4

Numbers

Numbers hold some kind of fascination for most people. They have been studied extensively for centuries and still there are simple questions about ordinary numbers to which no one knows the answers. Some of these problems are crucial to whole branches of mathematics while others appear to be mere curiosities of no consequence. I will introduce a sample of these questions later in the chapter.

The difficulty in studying numbers is that there are infinitely many of them and they are all different. This may sound a trite observation but I will show you a simple example of the trouble that it causes. The number 12 is *abundant*, meaning that the sum of its factors (not including 12) is actually greater than itself: $1 + 2 + 3 + 4 + 6 = 16$. Are there any odd abundant numbers? A little experimentation with small odd numbers might convince you that the answer is 'no'. You could then spend any amount of time looking for a proof of this conjecture and you would never find one, because there *are* odd abundant numbers—you just need to look a bit harder than you might expect in order to find them. From memory, I think the first is 945, which has as the sum of its factors

$$1 + 3 + 5 + 7 + 9 + 15 + 21 + 27 + 35 + 45 + 63$$
$$+ 105 + 135 + 189 + 315 = 975.$$

To this day no one knows whether or not there are any odd *perfect* numbers, that is to say an odd number that exactly equals the sum of its factors. There are perfect even numbers such as 6,

28, and 496 as you can check for yourself. The even perfects are well understood in, so far as it was proved by Euler in the eighteenth century that they are in one-to-one correspondence with the so-called *Mersenne primes*: that is prime numbers of the form $2^p - 1$ where p is itself a prime. Given a Mersenne prime, an even perfect number can be found and Euler showed that every even perfect number arises in this way. The link between perfect numbers and Mersenne primes goes right back to Euclid, yet no one knows whether or not there are infinitely many primes of this special kind.

As you may realize, the structure of numbers is bound up with the prime numbers, so this shall be our starting point. Our subject for this chapter is the set of positive counting numbers, $\{1, 2, 3, \ldots\}$. Even 0 is not to be allowed into our discussion except by express invitation.

A number p is a *prime* if it has exactly two factors, which necessarily are the number p itself together with 1. The number 1 is not reckoned among the primes since it has but one factor, and so the first few primes are 2, 3, 5, 7, 11, 13, 17, . . . A number with more than two factors is called *composite*.

The primes are the multiplicative building blocks of the counting numbers as it is clear that any number is either prime or can be broken down into a product of primes; for instance, $60 = 6 \times 10 = 2 \times 3 \times 2 \times 5$. We say that $2^2 \times 3 \times 5$ is the *prime factorization* of 60. How do we know that there is no other? Perhaps it is possible to factorize some numbers as products of primes in several completely different ways. I find that most people are sure that this is not the case and indeed feel a little affronted by so wild a suggestion: if numbers were liable to behave this badly, they surely would have heard about it by now. This is quite right, but uniqueness of prime factorization is not obvious even though it may be familiar. It hinges on the following characteristic property of prime numbers.

Euclid's Lemma. If the prime p is a factor of the product ab then p is a factor of a or a factor of b (or perhaps a factor of both).

Composite numbers do not have this property; for example, 6 is a factor of $8 \times 9 = 72$ but is not a factor of either of the numbers 8 or 9. I used Euclid's Lemma in a slightly sly way in the second chapter, where I ran through the argument that $\sqrt{2}$ is irrational. I said there that if 2 is a factor of a^2 then a itself must be even. This follows from Euclid's Lemma upon taking $p = 2$, the only even prime, and taking $b = a$. Indeed, using Euclid's Lemma it is not hard to generalize the argument showing $\sqrt{2}$ to be irrational to prove that $\sqrt{p}$ is irrational for any prime p.

If we take Euclid's Lemma for granted, it is easy to convince ourselves that it is impossible to have four different prime numbers p, q, r, s with the property that $pq = rs$. Suppose that in fact this could happen. Since p is a factor of pq, it is equally a factor of rs; then by Euclid we can conclude that p is a factor of one of the primes r or s. Let us suppose it is r. However, a number greater than 1 can be a factor of r only if it is equal to r since r is prime: hence $p = r$, and we can cancel the common factor in the equation $pq = ps$ to yield $q = s$ also. Therefore the two prime factorizations have turned out to be the same after all. This can be generalized to any number of prime factors without real difficulty: suppose $p_1 p_2 \ldots p_n = q_1 q_2 \ldots q_m$ where all the ps and qs are primes, and suppose, for convenience, that we have arranged both the ps and qs in increasing order. (This does not preclude the possibility that two or more of the ps are equal, and the same goes for the qs.) Using Euclid's Lemma, we can conclude as before that $p_1 = q_1$. We then cancel and repeat the argument $n - 1$ times to yield the conclusion that n, the number of prime factors on the left, must exactly match m, the number of primes on the right, and that $p_1 = q_1$, $p_2 = q_2$, $\ldots$, etc.

We conclude that, provided Euclid's Lemma is correct, there is only one way to factor any number as a product of primes.

A little later I will prove Euclid's Lemma in an unexpected way, through the use of the *Euclidean Algorithm* for calculating the highest common factor of two numbers. Before I venture on to that topic I shall settle one more question. There are certainly infinitely many prime numbers.

This is not as obvious as you might first think. It is not enough to argue that, since there are infinitely many numbers and each is a product of primes, there must be infinitely many primes. After all, there are infinitely many powers of two, e.g. 2, 4, 8, 16, 32, $\ldots$, but there is only one prime, namely the prime 2, involved in the factorization of all of them. It therefore does not immediately follow that there are infinitely many primes. Conceivably, there could be just a fixed number of primes—ten say, with every number being a product of these ten primes—although of course very large numbers would involve high powers of some of these primes. I am sure that you still believe that nothing of the kind is true, but since there is no infinite list of prime numbers that we can turn to, how can we be sure that the primes do not simply run out after a time? We know because of the following simple argument found in Euclid.

Let $p_1, p_2, \ldots, p_n$ denote the first n primes; for example, if n were 10 this list would be 2, 3, 5, 7, 11, 13, 17, 19, 23, 29. Put N equal to the product of the first n primes and consider $N + 1 = p_1 p_2 \ldots p_n + 1$. Now every number, including $N + 1$, has some prime factor. However, for each of the primes p in our list, $\frac{N}{p}$ is a whole number (as p is a factor of N), so it follows that $\frac{N+1}{p} = \frac{N}{p} + \frac{1}{p}$ is not a whole number. We conclude that, although $N + 1$ has at least one prime factor, it cannot be any of the primes $p_1, p_2, \ldots, p_n$ and so must be larger than all of them. It follows that there is some prime number q, dividing $N + 1$, such that $p_n < q \leq N + 1$. In particular, this shows that, given any such list of primes, $p_1, p_2, \ldots, p_n$, there is always at least one more prime not in that list and therefore the totality of all prime numbers must be infinite.

Finding common factors by subtraction

In school we were all taught about the highest common factor (h.c.f.) d of two numbers a and b. For example the h.c.f. of $a = 12$ and $b = 8$ is $d = 4$. The idea arises when searching for the lowest common denominator in order to do an addition involving two fractions. If the denominators in question are a and b, then the lowest common denominator is the least common multiple of a and b, which can be shown to be equal to $\frac{ab}{d}$. In our example above this gives $\frac{(12 \times 8)}{4} = 12 \times 2 = 24$.

How do we find d? I do not personally recall being shown how, although d can be expressed simply enough in terms of the prime factorizations of a and b. For example, if $a = 2058 = 2 \times 3 \times 7^3$ and $b = 3675 = 3 \times 5^2 \times 7^2$, then $d = 3 \times 7^2 = 147$: the prime factors of d are those primes common to both a and b, and the power to which each such prime occurs in the factorization of d is the *minimum* of the two powers of that prime in the factorizations of a and b. Although this solves the problem, it involves much more work than is necessary, however. It is possible to find d without factorizing either a or b, and this is important as, in general, it is extremely difficult to find the prime factors of large numbers even though, in principle, it can always be done by trial and error.

The process for finding the highest common factor d of two numbers a and b is called the Euclidean Algorithm, which runs as follows.

1. Subtract the smaller number from the larger.
2. Discard the larger, and repeat step 1 with the two remaining numbers.
3. Continue until the two remaining numbers are equal; this final number is d.

Let us apply the algorithm to the pair $(3675, 2058)$. The number pairs we obtain run as follows:

$$(3675, 2058) \to (2058, 1617) \to (1617, 441) \to (1176, 441)$$
$$\to (735, 441) \to (441, 294) \to (294, 147) \to (147, 147);$$

and so in this case $d = 147$, as we have already calculated through the prime factorizations. So we have found d without factorizing the given numbers 2058 and 3675. Notice that the maximum of the two numbers in each pair decreases, those maxima being:

$$3675, 2058, 1617, 1176, 735, 441, 294, 147.$$

Euclid's is perhaps the earliest example of a true *algorithm*—a mechanical procedure for deciding a question. In this age of computing, I would have thought this would be an attractive topic for secondary schools to rediscover. Why does it work?

Perhaps the first question a computer scientist asks about an algorithm is 'does it halt'? The procedure takes you into a loop which you will travel around a number of times—we certainly do not want to be caught in the loop for ever. However, it really must stop—we begin with two positive numbers and each time we go through step 2 the maximum of the two numbers certainly decreases. This cannot go on for ever, for eventually this maximum would decrease to 0. This happens if and only if the two numbers in hand at some stage are identical (look again at the worked example), at which point we are told to stop. Hence the algorithm does produce a number, but why is it necessarily the highest common factor d?

The reason is that the procedure preserves all common factors of the two numbers at each stage, as I will explain. Suppose we begin with a and b and that a is the larger of the two numbers. We then perform the first step, $a - b = r$ say, and we proceed with the pair b and r instead. Suppose that c is any common factor of a and b, so that $a = cx$, $b = cy$ say. Then $r = cx - cy = c(x - y)$, so that c is also a factor of r. In the same way you can check (and I suggest you do), using the equation $a = b + r$, that any common factor of b and r is also a factor of a. We conclude that the set of common factors of a and b is the same as the set of common factors of b and r. In particular, the highest member of this set of common factors, the h.c.f. of a and b, is equally the h.c.f. of b and r. Each time we go through the loop, although the pair of numbers in hand changes, the h.c.f. of the pair is always the same. Eventually, as we have already seen, the two numbers are the same, and the highest common factor of two equal numbers is that number itself.

There are two more important things I wish to say about the Euclidean Algorithm, one practical, the other theoretical. Suppose we were to perform the algorithm on the pair $(92, 8)$. The step-by-step approach would have us subtracting 8 from 92 many times:

$$(92, 8) \to (84, 8) \to (76, 8) \to \ldots$$

The number of times we would subtract 8 in this way is 11, the number of times that 8 will go into 92. Clearly, we can speed things up by dividing 92 by 8 and subtracting that many multiples of 8 in one step:

$$92 = 11 \times 8 + 4.$$

What this tells us is that we will go through the loop 11 times before 8 emerges as the maximum of the two numbers in hand and at this stage the remainder will be 4. We then have $8 = 2 \times 4$ with remainder 0, indicating that after going through the loop twice more the remainder will be 0. This shows that it was at the previous pass through the loop that both numbers in hand were equal, and they were equal to 4, the highest common factor of 92 and 8:

$$(92, 8) \to \ldots \to (8, 4) \to (4, 4) : d = 4.$$

In practice the algorithm usually is carried out in this fashion, so a typical calculation is as follows (at each stage we underline the two numbers in hand):

Find the highest common factor of $\underline{516}$ and $\underline{432}$:

$$516 = 1 \times \underline{432} + \underline{84}$$
$$432 = 5 \times \underline{84} + \underline{12}$$
$$84 = 7 \times \underline{12}.$$

Since the remainder is now 0, the required highest common factor is 12.

The theoretical point is that we can use these equations to express the h.c.f., in this case 12, in terms of the original pair of numbers as follows. We begin with the second last equation and write:

$$12 = 432 - 5 \times 84. \tag{1}$$

Now we can use the first equation to express the intermediate remainder 84 in terms of 516 and 432:

$$84 = 516 - 432. \tag{2}$$

Substituting (2) into (1) then gives:

$$12 = 432 - 5(516 - 432) = 432 - 5 \times 516 + 5 \times 432;$$

that is,

$$12 = 6 \times 432 - 5 \times 516. \tag{3}$$

There are two things to notice. First, although we intended to deal only with positive numbers, we have been led, against our will as it were, to multiplication of negatives by negatives: $-5 \times -432 = 5 \times 432$. If this worries you, rest assured that we will return to this topic in the next chapter. At this stage we need only note that it does happen, and that the final equation (3) is in fact true, as you can check directly for yourself.

Now to the second point. Working the equations of the Euclidean Algorithm backwards shows that it is always possible to find integers x and y such that $d = ax + by$, although they are not necessarily positive, as we have seen in the previous example: $a = 516$, $b = 432$, and it transpired that $d = 12$, $x = -5$, and $y = 6$. Suppose that the numbers a and b were *coprime*, which is just a word meaning that their h.c.f. is 1. (For example, the pair 40 and 21 are coprime whereas the pair 24 and 21 are not coprime as the h.c.f. of the latter pair is 3.) Then there would exist integers x and y such that $ax + by = 1$. It is the fact that we can express the number 1 in this way that now allows us to return to complete the demonstration of Euclid's Lemma, which you will recall said that, if a prime p divides a product ab, then it in fact divides at least one of the numbers a or b.

Suppose that p is prime and is a factor of the product ab, so that $ab = rp$. Suppose that p is not a factor of a. (If it is, of course, we have proved our point.) Then, since p is prime, the h.c.f. of the pair a and p must be 1, so by the Euclidean Algorithm there are integers x and y such that $ax + py = 1$. Now:

$$b = b \times 1 = b(ax + py) = bax + bpy.$$

Since $ba = pr$, we substitute accordingly to obtain:

$$b = prx + pby = p(rx + by).$$

But this shows that p is a factor of b, which is exactly what we needed to prove. Thus Euclid's Lemma is proved.

One final comment: the force of a mathematical result is not always apparent. Euclid's Algorithm allows us to write the highest common factor d in the form $ax + by$. At first sight, there may seem to be no reason why you should ever want to do this. It was, however, the fact that it was *possible* to write 1 in the form

$ax + py$ in the previous argument which allowed us to prove Euclid's Lemma.

Some curiosities, old and new

Since this is Mathematics for the Curious, I shall close with a few of the famous unsolved, or at least hideously difficult, problems about numbers.

The Goldbach Conjecture

In the eighteenth century Goldbach conjectured that every even number greater than 2 is the sum of two primes. This seems to be true, and often there are many ways of showing it; for example, $18 = 11 + 7 = 13 + 5$. Try a few examples for yourself. Some weaker versions of this conjecture have been proved, but the original conjecture remains unproved. Not being a mathematician who specializes in number theory, I do not feel qualified to judge if the Goldbach Conjecture is considered a 'serious' problem. I have heard it dismissed with the remark: 'Prime numbers were never meant to be added up.' Perhaps not, but one can detect an element of frustration in this kind of response.

Fermat's Last Theorem

One problem that certainly was taken very seriously was Fermat's Last Theorem. This requires a little introduction.

It is possible to have a right-angled triangle the sides of which are whole numbers. For example, even the ancient Egyptians appreciated that a (3, 4, 5) triangle had a right angle—this of course follows from Pythagoras's Theorem, which we proved in Chapter 3, as $3^2 + 4^2 = 5^2$. We can generate more such Pythagorean triples by simply multiplying all the numbers involved by 2 or some other factor. However, a (6, 8, 10) triangle is similar to the (3, 4, 5), the former being exactly the same shape, just twice the size. The difference between the two triangles is merely a matter of scale. But there are some genuinely different Pythagorean triples, such as (5, 12, 13) and (8, 15, 17). We ask the question: is it possible to describe all the Pythagorean triples (a, b, c), where a, b, and c have no common factor other than 1? The answer is yes, and the details are as follows.

This is the recipe. Take any pair of coprime numbers m and n, $m > n$, with one of them even. Put $a = 2mn$, $b = m^2 - n^2$, and $c = m^2 + n^2$. Then (a, b, c) is a Pythagorean triple for which a, b, and c share no common factor. This much is easy to prove: certainly if your algebra is up to it you can check directly that $a^2 + b^2 = c^2$. The hard part is proving that the converse is also true: for any Pythagorean triple (a, b, c) with no common factor there are coprime integers, m and n, with one of them even, such that a, b, and c are given by the above formulae. However, all this has long been settled and I shall not record the details here, although they are not very difficult.

We now look at powers higher than 2. What Fermat asserted in the early seventeenth century was that it is impossible to find two cubes that sum to another cube, impossible to find two fourth powers that yield another when added, and so on. That is to say, for $n \geq 3$ there are *no integer solutions* to the equation:

$$x^n + y^n = z^n.$$

Fermat claimed to have a truly marvellous proof of this conjecture which appears as one of the marginal notes in his manuscripts. He said that the margin was too small to contain the proof, and so he never seems to have written it down. Fermat made a number of similar marginal notes, all of which had been resolved satisfactorily except this one—hence the title of Fermat's Last Theorem.

At first sight this may not look a particularly interesting question, but such a superficial judgement would be utterly wrong, as more wonderful mathematics has flowed from the study of Fermat's Last Theorem than any other question. Happily, the problem was solved, in the way Fermat predicted, by Andrew Wiles in the 1990s, through proving an extremely deep conjecture on so-called elliptic curves and modular forms which has no obvious connection with Fermat. The proof, which has been hailed as the proof of the century, is so extraordinarily deep that only a handful of privileged people can claim to understand it fully. An early version of the proof, thought to have been complete, turned out to have a fundamental error, which was later triumphantly resolved by Wiles. Wiles's work certainly represents a wonderfully inspiring effort of human genius, even if it can be admired only from afar.

It is a shame that Andrew Wiles will not have his achievement honoured through the award of a Nobel Prize. There is no Nobel Prize for Mathematics.

And what of Fermat's original proof? The proof of Wiles is based on an enormous raft of nineteenth- and twentieth-century mathematics so was certainly outside the realms of anything Fermat could have devised in his own lifetime. Whatever Pierre de Fermat was thinking when he scribbled his marginal message remains a mystery, perhaps for ever.

Formulae for primes

I do not advise the reader to indulge in the quest for finding one of these, although it is natural for anyone to search for patterns among the primes. By a formula for primes we mean some kind of function, $f(n)$, such that, for any natural number n, $f(n)$ is the nth prime. Such a rule must of course begin with the values:

$$f(1) = 2, \ f(2) = 3, \ f(3) = 5, \ldots, \ f(10) = 29, \ldots$$

In a strict (and useless) sense, we know what $f(n)$ is: it is the value of the nth prime number. The trouble is that, in general, we have no simple way of calculating $f(n)$. We could begin with the more modest aim of insisting that, for each n, our function value $f(n)$ was a prime greater than the prime $f(n-1)$. In other words, we would settle for a formula that produced an increasing sequence of primes, even if some were missed out.

Certainly a formula like $f(n) = 6n + 1$ cannot work. (This first fails when $n = 4$.) Take any formula of the form $f(n) = an + b$, where a and b are integers. It is hopeless to take $a = \pm 1$ as such a formula gives every number from b onwards if $a = 1$ and every number from b downwards if $a = -1$; so let us suppose otherwise $a \neq \pm 1$. Then $f(b) = ab + b = b(a + 1)$, which is thus a composite number unless $b = \pm 1$ when it might yet be prime. Suppose then that $b = 1$. Then we can get a composite number out by putting $n = a + 2$, since:

$$f(a + 2) = a(a + 2) + 1 = a^2 + 2a + 1 = (a + 1)^2,$$

which, being a square, is composite. (For instance, if $a = 6$ we get $f(8) = 6 \times 8 + 1 = 49 = 7^2$.) If we try $b = -1$ we meet trouble upon putting $n = a$: $f(a) = a^2 - 1 = (a + 1)(a - 1)$. (For example, if $a = 6$ we get $f(6) = 6 \times 6 - 1 = 35 = 7 \times 5$.)

You may be tempted to try formulae involving higher powers of n such as $f(n) = n^2 + n + 41$, but it is always possible, by using a slightly more sophisticated approach to that above, to find an input that yields an output which has proper factors. This quadratic example, due to Euler, is nevertheless remarkable in that it does yield primes for 80 consecutive integers: $n = -40$, $-39, \ldots, 39$. For example, for $n = 7$ we obtain the prime 97. You may care to try a few other values of n. For 40, however, it fails as $f(40) = 1681 = 41 \times 41$. You can see this coming if you use a little adept factorization:

$$f(40) = 40^2 + 40 + 41 = 40(40 + 1) + 41$$
$$= 40 \times 41 + 41 = (40 + 1)41 = 41^2.$$

Let us look at some other questions on the topic of primes. It is possible to find arbitrarily long strings of consecutive numbers that are free of primes. One proof involves the use of factorials. The number $n!$ (read *n-factorial*) is simply the product

$$n \times (n - 1) \times (n - 2) \times (n - 3) \times \ldots \times 3 \times 2 \times 1$$

(although we can leave off the $\times 1$ as it clearly makes no difference). The factorials grow very quickly—$6! = 6 \times 5 \times 4 \times 3 \times 2 = 720$—and they soon reach many billions. They arise constantly in problems involving counting and probability, such as finding the chances of winning the National Lottery (about 1 in 14,000,000: see Chapter 6). We use the fact that $n!$ has many factors, including all of the numbers 2, 3, ..., n, to construct a sequence of n consecutive composite numbers for any given n. Consider the following numbers:

$$(n + 1)! + 2, \ (n + 1)! + 3, \ldots, (n + 1)! + n + 1.$$

There are n consecutive numbers here. The first is divisible by 2, because both the terms $(n + 1)!$ and 2 are even; the second is divisible by 3, as 3 is also a factor of $(n + 1)!$ and is undeniably a factor of 3; and so on, the last number having $n + 1$ as a factor. In particular, none of these numbers are primes. The sequence therefore represents a list of n consecutive composite numbers.

This sequence also serves to show that no sequence of the form $an + b$ can consist only of primes because the common difference between consecutive terms in such a sequence is always a, but we have just seen that the gap between successive primes can be made as large as you please.

A lot is known about the overall frequency of prime numbers. There is always at least one prime p such that $n < p < 2n$ for any $n \geq 2$; and $\frac{n}{p(n)}$, where $p(n)$ denotes the number of primes less than or equal to n, is a ratio that in the long run is known to approach $\log_e n$, the so-called *natural logarithm* of n.

There are formulae for primes: one is a rather elaborate reformulation of the problem which yields a formula that, though it looks impressive, could be used only if we knew what all the primes were in the first place. Another is a genuine formula, but the amount of calculation required to use it is so fantastic that it does not do the job either.

Since there is no useful formula for primes, at *any one time* there is a largest known prime. The champion at any stage is usually a *Mersenne prime*, that is a prime of the form $2^p - 1$ where p is also a prime. Although it is known that Mersenne numbers are not always prime ($2^{67} - 1$ is divisible by $193, 707, 221$, for example), it can be proved that any divisor of a Mersenne number has the form $2kp + 1$ for some positive integer k. This makes them, to use the word very precisely, prime candidates. For instance, let us check the Mersenne number $2^{11} - 1 = 2047$ for primality.

First, let me make the point that any composite number n must have a factor that is no more than $\sqrt{n}$ because factors come in pairs, and it is impossible for both factors to exceed $\sqrt{n}$ for then their product would exceed n. For example, $77 = 7 \times 11$, and 7 is a factor of 77 less than $\sqrt{77}$, which lies between 8 and 9. Since any factor itself has a prime factor, it follows that, in order to show that a number n is prime, it is enough to verify that it has no prime factor less than or equal to $\sqrt{n}$. In order to verify whether or not $2047 = 2^{11} - 1$ is prime, therefore, we need only check for divisibility by primes of the form $22k + 1$ which are themselves less than 46, as $46^2 > 2047$. There is only one such prime: putting $k = 1$ gives 23 which is a possible prime factor. Indeed it is, as division by 23 gives $2047 = 23 \times 89$. In fact 2047 is the first composite Mersenne number. (Note that 89 also has the form $22k + 1$.)

Even the relatively large Mersenne number $M = 2^{19} - 1 = 524{,}287$ can be shown to be prime by hand calculation. This time we begin by noting that $725^2 > M$, so only primes of the form $38k + 1$ that do not exceed 724 need be checked. Only the values $k = 5$, 6, 11, 12, 15, and 17 yield such primes, giving six divisions to be carried out.

Not all simple unsolved problems about numbers are old. Recently it was noticed that, beginning with any natural number n, the following process always seems to end with the number 1. If n is even, divide it by 2; if n is odd, multiply by 3 and add 1. For example, beginning with 7 we are led by the rules through the following sequence:

$$7 \to 22 \to 11 \to 34 \to 17 \to 52 \to 26 \to 13 \to 40$$
$$\to 20 \to 10 \to 5 \to 16 \to 8 \to 4 \to 2 \to 1.$$

No doubt this conjecture has been checked for all values of n up to some extraordinarily high number, but no one has yet come up with a reason why it must happen every time.

Pascal's Triangle and counting subsets

A very fundamental type of number goes by the name of *binomial coefficient*. The reason for this title will be explained in the next chapter. For the present we shall use the term *binomial number*, $C(n, r)$, for the number of different sets of r objects that can be chosen from a collection of n objects. These numbers turn out to be so amenable to manipulation that they often can be used to settle what would otherwise be awkward questions. For instance, the product of any four consecutive integers is always a multiple of 4! = 24, e.g.:

$$17 \times 18 \times 19 \times 20 = 24 \times 4845.$$

The product of five numbers in a row is similarly a multiple of 5! = 120, and in general $r!$ is always a factor of any r consecutive integers. That this is true will be quite clear once we have learnt a little about the genesis of the binomial numbers.

As I shall explain, all the binomial numbers are displayed in the array known as Pascal's Triangle (Figure 1); we shall see that the nth row, read from left to right, consists of the numbers C(n, 0), C(n, 1), ..., C(n, r), ..., C(n, n). I shall explain how to generate the rows of the triangle in a moment, although you may like to try to discover the pattern yourself.

Pascal's Triange

```
            1  1
           1  2  1
          1  3  3  1
         1  4  6  4  1
       1  5  10  10  5  1
      1  6  15  20  15  6  1
    1  7  21  35  35  21  7  1
 1  8  28  56  70  56  28  8  1
```

•

•

•

Fig. 1

We can easily check the first few lines of the triangle by inspection. For example, the 6 in the centre of the fourth row says that C(4, 2)=6, which is true as there are six ways of choosing a pair of people from a set of four {A,B,C,D}, the six (unordered) pairs being AB, AC, AD, BC, BD, and CD.

An obvious symmetry of Pascal's Triangle is that each row is the same read backwards or forwards. In terms of the binomial numbers, this is saying that $C(n, r) = C(n, n - r)$; for example, the two instances of 56 in the final displayed row are saying that $C(8, 3) = C(8, 5)$. However, this is not surprsing when you think about it: whenever you choose three people from a set of eight you are at the same time choosing a set of five people by default—the five that you leave behind. It follows that the number of ways of choosing three from eight is the same as the number of ways of choosing five from eight.

What is the rule for writing down the next line of the triangle? Each entry of a line (apart from those on the ends, which are always 1) is found by adding the two immediately above. For example, the first 28 in line eight comes from $7 + 21$. What is this saying about the binomial numbers? It is saying that $C(8, 3) = C(7, 2) + C(7, 3)$, and in general:

$$C(n, r) = C(n - 1, r - 1) + C(n - 1, r).$$

If we could explain why this is the case, we could conclude that the Pascal Triangle really does give all the binomial numbers. Let us look at the case of $C(8, 3)$, although what I am about to say applies equally well to the general case. To make matters simple, let our eight-member set be $A = \{1, 2, 3, \ldots, 8\}$. The sets of three that we can choose from A are of two distinct types: those that contain 8 and those that do not. To choose a set of the first type, we take the number 8 and then are left to choose two numbers from 1 to 7: by definition there are $C(7, 2)$ ways of doing this. On the other hand, if we do not take 8 to be one of our three selections, we make a choice of three numbers from the first seven, which can be done in $C(7, 3)$ ways. Adding the two possibilities together, we obtain the stated result that $C(8, 3) = C(7, 2) + C(7, 3)$. This argument applies equally well to the general case—one just runs through it replacing 8 by n and 3 by r throughout.

The Pascal Triangle provides a method of calculating any given binomial number $C(n, r)$, although we first have to find entries in all previous rows. It would be nice to have a formula for $C(n, r)$, that is to say an expression for the number in terms of

n and r only. A direct attack on the question does yield this reward. Again let us consider C(8, 3).

The number of ways of choosing three objects *in order* from eight is $8 \times 7 \times 6$: the same object cannot be chosen twice so the number of possibilities for the next object drops by one after each selection. Each set of three objects covers $3 \times 2 \times 1$ orderings so the number of different sets of three is:

$$\frac{8 \times 7 \times 6}{3 \times 2} = 8 \times 7 = 56.$$

Applying this reasoning in general allows us to infer that C(n, r) is equal to the product of the r consecutive integers from n downwards divided by r! In other words:

$$C(n,r) = \frac{n(n-1)(n-2)\dots(n-r+1)}{r!}. \tag{4}$$

Note that the last term in the numerator is $n - r + 1$ and not $n - r$ because the first term is $n = n - 0$ and not $n - 1$; for example, for $n = 8$ and $r = 3$ the final term is $8 - 3 + 1 = 6$, as we have already seen.

Since we are free to take n to be any number greater than or equal to r, the numerator in expression (4) can be made to stand for the product of any r consecutive positive numbers. Since C(n, r) is undoubtedly a whole number and not a fraction, it follows that the product of any r consecutive integers is divisible by r!. For example, the product $11 \times 12 \times 13 \times 14 \times 15$ is a multiple of $5! = 120$, as putting $n = 15$ and $r = 5$ in (4) gives:

$$C(15, 5) = \frac{15 \times 14 \times 13 \times 12 \times 11}{5!}.$$

A more compact expression for (4) is obtained by noting that the numerator is equal to $\frac{n!}{(n-r)!}$. Again, taking the example where $n = 8$ and $r = 3$, we are saying that:

$$8 \times 7 \times 6 = \frac{8!}{(8-3)!} = \frac{8!}{5!},$$

which can be seen immediately through cancellation. This gives a standard expression for $C(n, r)$:

$$C(n,r) = \frac{n!}{r!(n-r)!}. \tag{5}$$

This form of the binomial number also makes it plain that $C(n, r) = C(n, n - r)$, as replacing r by $n - r$ in the right-hand side of (5) returns the same expression since $n - (n - r) = n - n + r = r$.

5

Algebra

It has been said of certain Ethiopian tribesmen that the only multiplicative operations they would allow were those of doubling and halving, and what is more they would have no truck with fractions of any kind. None the less, they had no trouble multiplying any two numbers together and so conducting basic commerce. For instance, if one bought 31 sheep from another at a cost of £25 each, here is how they would find the total cost.

Form two columns headed by the numbers 25 and 31. Double the figure on the right while halving the figure on the left, ignoring any remainders of $\frac{1}{2}$ that arise through halving an odd number. Continue doing this until the number on the left has been reduced to 1:

$$
\begin{array}{rr}
25 & 31 \\
12 & 62 \\
6 & 124 \\
3 & 248 \\
1 & 496 \\
\end{array}
$$

Cross out the rows containing even numbers in the left-hand column, in this case the two rows beginning with 12 and 6. Now sum the remaining numbers in the right-hand column to give $31 + 248 + 496 = 775$, the right answer.

If this African technique of multiplication seems obscure to us, no doubt ours would also seem so to them. Can you explain their method? Can you explain your own method? In fact both rely on the same idea, the same facet of algebra, known as the Distributive Law, which is the main theme of this chapter. We shall return to the problem of the Ethiopian traders a little later.

I would like to begin, however, with a word about bracketing. When we write $2 + 4 + 7$ we do not feel the need for brackets as the two alternative ways of computing the sum lead to one and the same answer:

$$(2 + 4) + 7 = 6 + 7 = 13; \quad 2 + (4 + 7) = 2 + 11 = 13.$$

We say that the operation of addition is *associative*. The same applies to multiplication in that, for any three numbers a, b, and c, we have $(ab)c = a(bc)$.

When these operations are mixed, however, the bracketing matters:

$$2 + (4 \times 7) = 2 + 28 = 30; \quad (2 + 4) \times 7 = 6 \times 7 = 42.$$

What then does $2 + 4 \times 7$ mean? This expression would be inherently ambiguous if it were not for the fact that there is a *convention* (made by Man, not by God) that multiplication takes precedence over addition, so that $2 + 4 \times 7$ implicitly means $2 + (4 \times 7)$. If you intend the addition to be first, however, you need to communicate this through the use of brackets: $(2 + 4) \times 7$.

All of us have had some experience of tricky-looking algebra or arithmetic from our schooldays: you always had to beware when expressions involving minus signs and brackets were about. In the main this is due to the operation of subtraction *not* being associative. When performing two successive subtractions, the bracketing does matter:

$$9 - (4 - 2) = 9 - 2 = 7, \quad (9 - 4) - 2 = 5 - 2 = 3.$$

Again, the convention is that $9 - 4 - 2$ with no brackets implicitly means $(9 - 4) - 2$; in other words, the quantities are subtracted in the order in which they occur. The chances of misinterpretation are high, however, and for that reason you will often see the first pair of brackets written explicitly just to be on the safe side. The same applies to division: the operation is not associative and so the brackets are not just optional extras:

$$(32 \div 8) \div 2 = 4 \div 2 = 2; \quad 32 \div (8 \div 2) = 32 \div 4 = 8.$$

Again, if we write $32 \div 8 \div 2$ we mean $(32 \div 8) \div 2$, but, for the same reason as before, it is just as well to put in the brackets for the sake of clarity. Since division is not associative, it is best to avoid writing fractions with two floors—it looks ugly and without the brackets the meaning is ambiguous:

$$\frac{\frac{32}{8}}{2} = 2; \quad \frac{32}{\frac{8}{2}} = 8.$$

The Distributive Law is the most peculiar and least obvious of all the laws of algebra. Its peculiarity lies in the fact that it is the only one that links the two main operations of arithmetic, addition and multiplication: it is the law that tells you how to multiply out brackets:

$$a(b + c) = ab + ac.$$

For example, $4(2 + 3) = 4 \times 2 + 4 \times 3$, which is certainly true as $4(2 + 3) = 4 \times 5 = 20 = 8 + 12$. A moment's reflection on an example such as this allows you to see what is going on. On one side of this sum we have:

$$4(2 + 3) = (2 + 3) + (2 + 3) + (2 + 3) + (2 + 3),$$

while the other is:

$$(2 + 2 + 2 + 2) + (3 + 3 + 3 + 3).$$

Both sums involve the same numbers but merely added in a different order, which of course makes no difference. (That is a law too: the *Commutative Law of Addition*: $a + b = b + a$.) The general case holds for the same reason, for it says:

$$\underbrace{(b + c) + (b + c) + \ldots + (b + c)}_{a \text{ times}} = \underbrace{(b + b + \ldots + b)}_{a \text{ times}}$$

$$+ \underbrace{(c + c + \ldots + c)}_{a \text{ times}}.$$

The subtlety in the Distributive Law comes from using it in the reverse direction to express a sum as a product by 'taking out' a common factor. Unlike expansion of brackets, an entirely mechanical process explained below, factorization involves considering an expression and spotting a common factor. It requires judgement on behalf of the student, who must have the confidence to decide that the factorization is appropriate and will help in the task of simplifying a certain algebraic expression. This requires considerable experience, but it is an unavoidable part of the learning process. Serious students of a mathematically based subject need to be at ease with their algebra and to know how to make use of the Distributive Law in both directions.

An example of this factorization process occurred on page 87 where we checked that $40^2 + 40 + 41 = 41^2$: there the Distributive Law was used twice. In fact, the Distributive Law is at work whenever we carry out an ordinary multiplication, however we may go about it. Our method relies on three things: knowing our tables up to 10 (the base of our number system), knowing that to multiply by 10 we simply adjoin a 0 to the end of the number to be multiplied, and the Distributive Law.

For example, multiply 32 by 7:

$$
\begin{array}{r}
32 \\
\times 7 \\
\hline
224
\end{array}
$$

You have actually carried out two little multiplications using your knowledge of the times table, then multiplied by 10, and finally completed the sum by adding the answers together. What the method entails *if written down explicitly* is the following:

$$32 \times 7 = (30 + 2) \times 7 = 30 \times 7 + 2 \times 7.$$

We have used the Distributive Law to split the multiplication into the sum of two simpler ones. Next, we say:

$$30 \times 7 + 2 \times 7 = 30 \times 7 + 14 = (30 \times 7 + 10) + 4.$$

The first step involves your knowledge of the 2 times table and the next step is called the *carry*, where you take the 10 that has arisen and carry it over into the tens column. The entry in the units column is now settled. We continue; the remaining steps are justified by the Commutative Law of Multiplication (numbers can be multiplied in any order), knowledge of the 3 times table, the Distributive Law again, our rule for multiplication by 10, and finally a simple addition:

$$= (3 \times 10 \times 7 + 10) + 4 = (3 \times 7 \times 10 + 10) + 4$$
$$= (21 \times 10 + 10) + 4 = (21 + 1) \times 10 + 4$$
$$= 22 \times 10 + 4 = 220 + 4 = 224.$$

You may feel a little uncomfortable with this detailed level of explanation. Part of the reason for this is that I am explaining something very familiar—simple multiplication—using ideas that may be less familiar—the laws of arithmetic. If that disturbs you, simply let it pass by while noting one general point: every arithmetic method depends for its justification on a handful of very simple laws of arithmetic, the Distributive Law being one of them. I hope nevertheless that the laws of arithmetic will help clarify the Ethiopian method of multiplication because, unless you are an Ethiopian tribesman, you will probably feel that it does stand in need of explanation.

Let us now return to the Ethiopian multiplication problem with all its doubling and halving, throwing away halves, crossing out even rows, and summing the rest. This may seem bewildering to us but, when analysed, it is all seen to be firmly supported by the Distributive Law.

First, let us look at an example where the African approach is transparent. The basic idea is to calculate ab by replacing it by $\frac{a}{2} \cdot 2b$. If one of the numbers is a power of 2, then the method is straightforward. For example, to calculate 16×40 the tribesmen would say:

$$16 \times 40 = 8 \times 80 = 4 \times 160 = 2 \times 320 = 1 \times 640 = 640.$$

In this case every row, except the final one, 1×640, begins with an even number so that all rows but this one are crossed out.

This is all clear enough. The obscure point of the method arises when we hit odd numbers in the left-hand column. Let us look carefully at this. Suppose that a row is $a\ b$ with a odd. What is effectively done at this stage is to write a as $c + 1$ (of course, $c = a - 1$) and expand using the Distributive Law:

$$ab = (c + 1)b = cb + b.$$

We then continue working out the product cb; the extra b cannot be forgotten, however, and that is why the entries in the right-hand column that begin with an *odd* number are not discarded but form part of the final sum. The Ethiopians appeared to be ignoring the fractions that arose in their calculation, but in fact their bookkeeping was sound. Let us go through the example given at the beginning of the chapter using our modern notation to make explicit the Ethiopian method:

$$25 \times 31 = (24 + 1) \times 31 = 24 \times 31 + 31 = 12 \times 62 + 31.$$

We see that, in passing from the first row to the second, the 25 is replaced by 24 and then the halving and doubling step is carried out, taking us to the second row; however, the 1×31 is not overlooked: it sits in the second column of the first row waiting to be collected later.

Continuing, we obtain:

$$= 6 \times 124 + 31 = 3 \times 248 + 31 = (2 + 1) \times 248 + 31$$
$$= 2 \times 248 + 248 + 31 = 496 + 248 + 31 = 775.$$

Although it may not be the approach with which we would feel comfortable, the Ethiopian Method is perfectly sound, as are many other methods of doing multiplication. This is an important psychological point. When doing mental arithmetic, everyone seems to have their own idiosyncratic methods. Provided that they work, there is *nothing wrong with this*. Often people are somewhat sheepish about owning up to having their own way of doing things because they fear being told that they are going about their sums the *wrong way*.

We may all have been encouraged to do mental arithmetic, but usually we are told little about how we are supposed to do it: we are largely left to our own devices. The standard methods for doing sums are designed to be used in conjunction with pencil and paper, where you have the advantage that numbers can be written down (carry-overs, for instance) and so stored without having to remember them as you pass to the next stage of the calculation. For that reason, they are not well suited for mental arithmetic as it is difficult to keep some numbers in mind while manipulating others in your head. When it comes to mental arithmetic you should feel free to do anything that works. If you should ever write down your own approach in order to convince yourself of its validity, you will see that, at bottom, your method relies on the same laws of arithmetic as do all the rest.

From arithmetic to algebra

Algebra involves doing arithmetic with unspecified numbers, denoted by symbols, instead of specific ones. This allows us to describe a general method in one fell swoop. On the one hand, this takes some getting used to, but on the other, since the laws of algebra must, by design, be the same as the laws of arithmetic, the manipulations involved are not new. This is why confidence in arithmetic leads to competence in algebra.

Let us introduce some standard algebra and see what can be done with it. Using the Distributive Law, we can expand an expression such as $(x + y)^2$. For the moment, let a stand for the number $x + y$. Then:

$$(x + y)^2 = (x + y)(x + y) = a(x + y) = ax + ay.$$

Now $ax = xa = x(x + y) = x^2 + xy$; similarly, $ay = ya = y(x + y) = yx + y^2$. Combining all this yields:

$$(x + y)^2 = ax + ay = x^2 + xy + yx + y^2 = x^2 + xy + xy + y^2$$
$$= x^2 + 2xy + y^2.$$

For emphasis, we repeat:

$$(x + y)^2 = x^2 + 2xy + y^2. \tag{1}$$

We met this before in Chapter 3, where we noted that $(a + b)^2 = a^2 + 2ab + b^2$ followed from some simple geometric considerations.

There are a number of stories that are based on (1). Apparently Leonhard Euler once wrote it on a blackboard as a proof of the existence of God, knowing that no one else in the room at the time would dare reveal their ignorance by arguing with him. Bertrand Russell became a first-rate mathematician but as a boy was forced to chant 'the square of a sum is equal to the sum of the squares increased by twice their product'; he admitted not having the slightest idea what this meant and only knew that if he got it wrong his tutor would throw things at him.

The way in which square roots respect multiplication and division but not addition and subtraction is one of the great sources of sorrow for students of algebra. We now have the opportunity to clarify this. Let a and b denote positive numbers so we have no trouble with the square roots of negative numbers in what follows. It is true that:

$$\sqrt{ab} = \sqrt{a}\sqrt{b}; \quad \sqrt{\frac{a}{b}} = \frac{\sqrt{a}}{\sqrt{b}}.$$

To see the first statement, we need only square both sides: the square of the left-hand side is, by definition, ab, while the right-hand side gives:

$$(\sqrt{a}\sqrt{b})(\sqrt{a}\sqrt{b}) = (\sqrt{a}\sqrt{a})(\sqrt{b}\sqrt{b}) = ab.$$

Here we have used the fact that the product of numbers can be rearranged in any order, the so-called *Commutative Law* of multiplication, to give the required result. Similarly, the reader can check that squaring both sides of the statement on quotients gives the tautology $\frac{a}{b} = \frac{a}{b}$ and this verifies the second assertion. We should note, however, that in both cases we are using the fact that, if two positive numbers x and y have equal squares so that $x^2 = y^2$, then the numbers themselves must be equal. This is certainly true but not does in general hold for arbitrary numbers; for example, $2^2 = (-2)^2 = 4$ but $2 \neq -2$. For this reason care must be taken when indulging in this kind of verification.

In words, what we have shown is that the square root of a product is the product of the square roots and that the square root of a quotient is the quotient of the square roots; that is to say, the operations of multiplication and extraction of square roots when done in either order yield the same result. This fact is often used to simplify square roots of larger numbers:

$$\sqrt{72} = \sqrt{36 \times 2} = \sqrt{36}\sqrt{2} = 6\sqrt{2}.$$

However, it is absolutely not the case that we can interchange addition and the taking of square roots, as is easily seen by example:

$$\sqrt{9+16} = \sqrt{25} = 5; \quad \sqrt{9} + \sqrt{16} = 3 + 4 = 7.$$

In fact, if a and b are positive numbers the square root of the sum is always less than the sum of square roots. This is because $(\sqrt{a+b})^2 = a + b$ while the square of the sum of square roots is more:

$$(\sqrt{a} + \sqrt{b})^2 = (\sqrt{a})^2 + (\sqrt{b})^2 + 2\sqrt{a}\sqrt{b}$$
$$= a + b + 2\sqrt{ab}.$$

We also are now in possession of enough algebra to derive the famous formula for solving quadratic equations, for it affords us the general technique called 'Completing the Square'. Let us begin with an example. Solve:

$$x^2 + 6x - 16 = 0.$$

Add 16 to both sides:

$$x^2 + 6x = 16.$$

Now think of $x^2 + 6x$ as $x^2 + 2xy$: clearly, then, $2xy = 6x$ and $y = \frac{6}{2} = 3$. If the left-hand side of the equation were $x^2 + 2xy + y^2$ we would be able to write it as $(x + y)^2$, and continue by taking square roots. Our next step therefore is to add $y^2 = 3^2 = 9$ to both sides:

$$x^2 + 6x + 9 = 16 + 9 = 25.$$

The left-hand side is now a perfect square, $(x + 3)^2$, and we can solve the equation without difficulty provided that we remember that a positive number has a negative square root as well as a positive one:

$$(x + 3)^2 = 25$$
$$\Rightarrow x + 3 = \pm\sqrt{25} = \pm 5,$$

where ± 5 means $+5$ or -5 and the symbol $\Rightarrow$ stands for 'implies'. Hence:

$$x = 5 - 3 = 2 \text{ or } x = -5 - 3 = -8.$$

In order to use algebra efficiently we need to be able to deal with both positive and negative numbers. The reason is that, even when dealing with problems involving only positive quantities and positive solutions, the algebraic operations may lead us out of the realm of positive numbers into that of negatives, although we may eventually return to the positive world. If we want to be able to divide freely we need fractions, and if we wish to subtract freely we require negative numbers as well as positive ones.

There seems not to have ever been much reluctance to use fractions—I suppose because the idea of a fraction of a physical object still makes sense, at least on some occasions. (Inappropriate fractional descriptions can be the butt of jokes, as in '2.4 children'.) As mentioned in Chapter 2, the ancient Egyptians restricted themselves to fractions with 1 in the numerator. This self-imposed handicap leads to some pretty problems that we shall take up again a little later.

There has always been a bias in favour of positive numbers, however. The Babylonians knew how to solve quadratic equations but would admit only positive solutions. For this reason their standard way of presenting a quadratic equation may look strange to us. The favoured approach was to ask for the dimensions of a rectangle given its perimeter and area. This ensured that positive solutions were always available.

This type of problem is equivalent to a single quadratic equation. Suppose the perimeter of the rectangle is 28 units and the area is 48. Let x and y denote the lengths of the sides. Then we have:

$$2(x + y) = 28, \quad xy = 48.$$

The second equation allows us to write $y = \frac{48}{x}$. Dividing both sides of the first equation by 2 and substituting for y then yields:

$$x + \frac{48}{x} = 14.$$

Multiplying all terms by x then gives:

$$x^2 + 48 = 14x \Rightarrow x^2 - 14x = -48. \tag{2}$$

We can now continue as before, completing the square, provided we know that, in general:

$$(x - y)^2 = x^2 - 2xy + y^2.$$

Accepting this for the moment, we add $\left(\frac{14}{2}\right)^2 = 7^2 = 49$ to both sides of (2) to get:

$$x^2 - 14x + 49 = -48 + 49 = 1$$
$$\Rightarrow (x - 7)^2 = 1.$$

Hence $x - 7 = \pm 1$ so $x = 7 + 1 = 8$ or $x = 7 - 1 = 6$. If $x = 8$ then $y = \frac{48}{8} = 6$, while if $x = 6$ then $y = \frac{48}{6} = 8$, so there is essentially a unique solution: an 8×6 rectangle.

Another example in which algebra reveals an interesting truth about positive quantities is the following. Once again, however, we shall allow ourselves to deal with negative numbers and use the fact that the product of two negative numbers is positive (a point that some people are very reluctant to accept).

There are several ways of averaging numbers. The common average of a and b is

$$\frac{a + b}{2}.$$

This is called the *arithmetic mean*. On the other hand, the *geometric mean* of two positive numbers is

$$g = \sqrt{ab}.$$

The geometric mean has the property that a square with side length g has the same area as the rectangle with sides a and b. It behaves like the ordinary average with respect to the taking of logs (although not important in what follows; for an explanation of logs see Chapter 2). Remembering that $x^{\frac{1}{2}}$ means $\sqrt{x}$, we can see this through use of the third and then first log laws:

$$\log g = \log(\sqrt{ab}) = \log((ab)^{1/2}) = \frac{1}{2}\log(ab) = \frac{\log a + \log b}{2}.$$

If you calculate a number of these means you will soon notice that the geometric mean is always less than the arithmetic. For example, if $a = 4$ and $b = 9$ the arithmetic mean is $\frac{(4+9)}{2} = 6.5$ while the geometric mean is $\sqrt{4 \times 9} = \sqrt{36} = 6$. Try a few yourself.

Now let's show that this is always true by using a little algebra. Let a and b be positive numbers and consider $(a - b)^2$. Now $a - b$ may be negative, but none the less the square of any number c is not negative. (In fact it is positive unless $c = 0$.) Using this and our expansion of $(a - b)^2$, we obtain:

$$(a - b)^2 \geq 0 \Rightarrow a^2 + b^2 - 2ab \geq 0.$$

Add $4ab$ to both sides of this inequality in order to make the left-hand side a perfect square, as shown:

$$a^2 + b^2 + 2ab \geq 4ab.$$

Hence

$$(a + b)^2 \geq 4ab.$$

Taking positive square roots of both sides now gives:

$$a + b \geq 2\sqrt{ab},$$

whereupon dividing by 2 gives the inequality:

$$\frac{a + b}{2} \geq \sqrt{ab},$$

which is what we want: the arithmetic mean is greater than or equal to the geometric mean.

In fact we can say more. If $a = b$ then both the arithmetic and geometric means are equal to a. If $a \neq b$ then $(a - b)^2 > 0$ and the argument above shows that the arithmetic mean is then strictly greater than the geometric.

If we are uninhibited with our algebra in making free use of negatives we can solve any quadratic equation by completing the square in the general case. For a general quadratic equation, i.e.

$$ax^2 + bx + c = 0,$$

completion of the square leads to the famous quadratic formula:

$$x = \frac{-b \pm \sqrt{b^2 - 4ac}}{2a}.$$

Although the algebra for deriving this result is a little tough by school standards, there is nothing new in it. The one fresh idea in solving quadratics is the completion of the square. Once that is appreciated, the general formula is straightforward. It does, however, require the student to have the right algebraic reflexes in order to be able to master this level of manipulation. For instance, in the course of the derivation, at one stage we are required to place a complicated algebraic expression over a common denominator. What is the justification for this?

All that is being carried out is the addition of fractions. The expressions may be complicated, but numerators and denominators still stand for (unspecified) numbers and so are subject to the same laws of algebra. Which laws are relevant? To answer this consider the addition of two fractions:

$$\frac{a}{b} + \frac{c}{d}.$$

The number bd is the product of both b and d so bd can act as a common denominator. We multiply the denominator b by d and so we multiply the numerator a by d as well. The overall effect is to multiply by $\frac{d}{d} = 1$ which does not change its value. Similarly we multiply $\frac{c}{d}$ by $\frac{b}{b}$ to give:

$$\frac{ad}{bd} + \frac{cb}{db}.$$

Now that our fractions have the same denominator, we may add the numerators to get:

$$\frac{ad}{bd} + \frac{bc}{bd} = \frac{ad + bc}{bd}.$$

This last stage uses the Distributive Law. To see it this way, ask yourself what is being said when we write something like:

$$\frac{x + y}{z} = \frac{x}{z} + \frac{y}{z}.$$

(In the above case, z is bd.) Division by z means multiplication by the reciprocal $s = \frac{1}{z}$. Hence this last statement becomes:

$$s(x + y) = sx + sy,$$

which is an instance of the Distributive Law with fractions.

The system of positive counting numbers is inadequate for algebra because two natural operations, subtraction and division, take us out of the system. Division leads to fractions, and the arithmetic of fractions is quite difficult but is readily accepted because it can be demonstrated explicitly using physical objects such as slices of pie, while subtraction leads to the negatives. Negative numbers took much longer to gain acceptance. Even during the Renaissance the validity of their use was questioned because they seemed to lack physical interpretation. Negative numbers are accepted as more meaningful in the modern world, perhaps because we are so familiar with the notion of debt, which is the interpretation of negative money. Of course, monetary debt has been around for millennia so this is not the whole story. Yet it is certainly true that people feel debt to be real, so the arithmetic of debt, which must at least allow for the addition of negative sums, is accepted. Debts grow larger when subject to interest. This involves multiplication of these negative values by positive multipliers to yield larger debt: if you have a £100 overdraft at 30% penalty interest, you will have an account worth £$(-100 \times 1.3) = -$£130 one year later.

A psychological sticking point, however, seems to be admitting that the product of two negative numbers is positive. This is exactly what you need in order that your algebra be consistent. A simple example illustrating the necessity of this rule is the following: $(2 - 1)(2 - 1) = 1 \times 1 = 1$. On the other hand, treating subtraction as adding the opposite, we also have:

$$1 = (2 - 1)(2 - 1) = (2 + (^-1))(2 + (^-1)).$$

Using the Distributive Law, we expand the brackets by multiplying each term in the first bracket by each in the second and sum them all, giving four terms in all. The calculation continues:

$$2(2 + (^-1)) + (^-1)(2 + (^-1))$$

$$= (2 \times 2) + (2 \times (^-1)) + ((^-1) \times 2) + ((^-1) \times (^-1))$$
$$= 4 + (^-2) + (^-2) + ((^-1) \times (^-1)) = 0 + ((^-1) \times (^-1))$$
$$= (^-1) \times (^-1).$$

Thus we are led to

$$(^-1) \times (^-1) = 1:$$

anything else will yield the wrong answer. None the less, whenever a teacher claims that the product of two negatives is positive, he or she is liable to meet with derisory remarks along the lines of 'You can't multiply two negative piles of money together to get a positive pile!' Although this may sound a convincing objection, it is itself nonsense as it makes no sense to multiply one *money pile* by another *money pile* in the first place, whether or not you consider the piles to be credits or debits. What the mathematics is saying is that, in any situation where it *does make sense* to multiply two negatives together, the answer will be positive.

The rules therefore need to be that the product of two numbers of the same sign is positive, while the product of numbers with opposite signs is negative. We can now expand any product of brackets using the Distributive Law and these rules. To expand a bracket involving a subtraction, we treat the minus as meaning addition of the opposite: we just need to realize that $(^-a)b = a(^-b) = ^-(ab)$, as each is the opposite of the product ab:

$$a(b - c) = a(b + (-c)) = ab + a(-c) = ab - ac.$$

For example:

$$(x - y)^2 = (x - y)(x - y) = x(x - y) - y(x - y)$$
$$= x^2 - xy - yx - y(-y).$$

Now subtracting $y(-y) = -y^2$ means adding its opposite, y^2, and so we obtain:

$$(x - y)^2 = x^2 - 2xy + y^2.$$

In a similar way we derive the expression for the *Difference of Two Squares*:

$$(x + y)(x - y) = x(x - y) + y(x - y)$$
$$= x^2 - xy + yx - y^2 = x^2 - y^2.$$

Again, it is the use of this in the reverse direction that often needs to be appreciated. The difference of two squares can be written as a product. For example:

$$n^2 - 1 = n^2 - 1^2 = (n + 1)(n - 1),$$

which you may recall arose in Problem 4 of the first chapter.

The next two higher powers of $x + y$ are given by the following expressions:

$$(x + y)^3 = x^3 + 3x^2y + 3xy^2 + y^3$$
$$(x + y)^4 = x^4 + 4x^3y + 6x^2y^2 + 4xy^3 + y^4.$$

It is not hard to describe the general expansion of $(x + y)^n$. It may seem daunting at first sight as there will be many terms when all the brackets are expanded, making it difficult to keep track. However, we ask: what does a typical term look like? It has the form $x^r y^s$, where $r + s = n$. The reason for this is that any term involves taking one symbol, either x or y, from each of the n brackets and so the total number of xs and ys in each term must be n in all. For example, in the expansion of $(x + y)^4$, one term results from choosing x from the first, third, and fourth brackets and y from the second, giving a contribution of $xyxx = x^3y$ to the overall expansion. It now comes down to counting how many times each type of term arises.

The sequences of coefficients in the two examples above are, respectively, (1,3,3,1) and (1,4,6,4,1). If you turn back to the picture of Pascal's Triangle in the previous chapter you will see that these represent the third and fourth rows of that array. That is why the numbers in the Triangle are called the *binomial coefficients*, as the nth row allows you to write down the expansion of the nth power of the binomial (meaning two-term) expression $x + y$. The reason why this works is also readily explained once we recall that $C(n, r)$ counts the number of ways of choosing r objects from n available. To get a term of the type $x^r y^{n-r}$ in the binomial expansion we must choose the x symbol from r of the n brackets available and the y symbol from the remaining $n - r$. The total number of ways of choosing the r brackets from n is, by its very meaning, $C(n, r)$, and so the coefficient of $x^r y^{n-r}$ in the binomial expansion of $(x + y)^n$ is the binomial number $C(n, r)$. This fact is called the *Binomial Theorem*.

Egyptian fractions revisited

You will recall in Chapter 2 how we said that it is always possible to express a proper fraction, $\frac{m}{n}$, as a sum of distinct unit fractions. For example:

$$\frac{6}{13} = \frac{1}{3} + \frac{1}{8} + \frac{1}{312}.$$

The suggested method was to subtract the largest reciprocal available at each stage and the claim was that the process would finish in no more than m steps. Let us see why this is so.

First, what is the largest reciprocal less than the proper fraction $\frac{m}{n}$? Only the case where m is at least 2 needs attention, and let us assume that we have cancelled the fraction down to lowest terms so that m and n have no common factor other than 1. Now since $m < n$ we can divide m into n. Let us suppose that m divides into n a total of k times with remainder r so that:

$$n = km + r, \ 1 \leq r \leq m - 1, \ 1 \leq k.$$

The value of k is at least 1 as $m < n$. The remainder r is at least 1 as n is not a multiple of m since the two numbers have a highest common factor of 1. The largest reciprocal less than $\frac{m}{n}$ is then $\frac{1}{(k+1)}$ because:

$$km < n = km + r < km + m = m(k + 1).$$

Taking reciprocals (which causes the inequalities to change direction) gives:

$$\frac{1}{km} > \frac{1}{n} > \frac{1}{m(k+1)},$$

whereupon multiplying through by m and writing the smallest number first gives:

$$\frac{1}{k+1} < \frac{m}{n} < \frac{1}{k}.$$

That is to say, $\frac{1}{(k+1)}$ is the largest unit fraction less than $\frac{m}{n}$ as the next largest, $\frac{1}{k}$, is too big.

Now we know exactly how to perform the calculation. For the example above where $m = 6$ and $n = 13$, we begin with $13 = 2 \times 6 + 1$ so that our first value of k is 2. We thus subtract $\frac{1}{(2+1)} = \frac{1}{3}$ to get:

$$\frac{6}{13} - \frac{1}{3} = \frac{6 \times 3 - 1 \times 13}{39} = \frac{5}{39}.$$

Next, $39 = 7 \times 5 + 4$ so that our second value of k is 7, giving $\frac{1}{8}$ as our next reciprocal to subtract:

$$\frac{5}{39} - \frac{1}{8} = \frac{5 \times 8 - 1 \times 39}{312} = \frac{1}{312}.$$

Therefore we have:

$$\frac{6}{13} = \frac{1}{3} + \frac{1}{8} + \frac{1}{312}.$$

To show that the process will always end in m steps or fewer requires some algebra. Let us carefully look at what happens at the first subtraction:

$$\frac{m}{n} - \frac{1}{k+1} = \frac{m(k+1) - n}{n(k+1)}.$$

Remembering that $n = mk + r$, we see this equals:

$$\frac{m(k+1) - (mk+r)}{n(k+1)} = \frac{mk + m - mk - r}{n(k+1)} = \frac{m-r}{n(k+1)}.$$

The key observation lies in what has happened to the numerator—it has decreased from m to $m - r$. Since r is positive, we conclude that the numerator of the next remainder is always less than the previous one. It follows that after $m - 1$ steps or fewer the process will produce a remainder that is itself a unit fraction and so we shall be finished. (This is an example of an inductive argument as was first introduced in the question concerning sharing vodka in Chapter 1: the principle is to reduce the general case to a previous case; in this instance we show how to pass from a general value of m to a lower value, $m - r$.)

It remains only to note that the next reciprocal subtracted will always be smaller than the previous one (as this will ensure that our unit fractions are all different). By the way it was selected, we see that the next reciprocal cannot be larger than its predecessor; nor could it be equal, as $\frac{2}{(k+1)}$ is larger than $\frac{m}{n}$ since:

$$\frac{m}{n} < \frac{1}{k} < \frac{2}{k+1}.$$

The final fraction inequality is easily verified, as the cross-multiplication rule of Chapter 2 says that it is equivalent to:

$$k + 1 < 2k \Leftrightarrow 1 < k,$$

and, as we have already noted, this is true, as the fraction $\frac{m}{n}$ is proper.

6

More Questions Answered

In most countries of the Western world the citizens have the opportunity of playing a state-run lottery. Britain joined this game late, but since 1994 nothing has united the country like the National Lottery, and so a book such as this one is obliged to settle the next question.

1. What are your chances of winning the lottery?

The lottery models are pretty similar across the globe. In Britain the basic game involves choosing a set of six numbered balls (in any order), and you win if your choice matches the machine's random selection of six balls from a collection numbered 1 to 49.

The number of ways six balls can emerge, taking into account the order, is $49 \times 48 \times 47 \times 46 \times 45 \times 44$: the same ball cannot be chosen twice, so the number of possibilities for the next ball drops by one each time. You have chosen a particular set of six balls which will cover $6 \times 5 \times 4 \times 3 \times 2 \times 1$ of these possible orderings. Therefore your chance of winning is:

$$\frac{6 \times 5 \times 4 \times 3 \times 2 \times 1}{49 \times 48 \times 47 \times 46 \times 45 \times 44} = \frac{1}{49 \times 47 \times 46 \times 44 \times 3}$$
$$= \frac{1}{13,983,816}.$$

You have one chance in 14 million. The lottery relies on our not appreciating just how enormous this number is. A line of 14 million biros could stretch from England to Mongolia—could you really expect yours to be the one chosen at random from among them?

The only advice a mathematician can give if you do decide to gamble on lotteries is as follows. First, pick high numbers, because small ones, in particular those under 32, are popular, presumably because people often use birthdays of their friends and relations. By selecting predominantly large numbers, should you win a big prize, it is likely to be very big, as you will have picked numbers that few other people have picked. Second, and for peace of mind this is the most important, don't pick the same numbers every time—if you do, you may feel obliged to play every week and spend the rest of your life in terror of your 'lucky' numbers coming up in the one week that you forget to enter! Certainly don't choose 1, 2, 3, 4, 5, 6—thousands of people do each week; should this combination ever come up, their portion of the shared jackpot will be tiny. Of course they will still have won more than those who picked other combinations. The point, however, is that you can *never* win a big prize with this or similar combinations as they are just too popular.

There is another, more dynamic, way of solving this question, more in keeping with the tension of the real situation. The chances that your particular choice of six will still be 'live' after the first ball rolls out is $\frac{6}{49}$ because you start with six of the 49 possible numbers. Of these lucky weeks, the chances that your choice is still 'live' after the second ball will be $\frac{5}{48}$, as the machine has 48 numbered balls left, and you hold five of the numbers, the sixth already having been used on the first ball. Hence the proportion of weeks when you are still in with a chance after two balls have been rolled is $\frac{5}{48}$ of $\frac{6}{49}$, that is the product

$$\frac{6}{49} \times \frac{5}{48}.$$

Continuing in this way, we see that the chance you have of staying 'live' after the full six balls have been drawn is, as before,

$$\frac{6}{49} \times \frac{5}{48} \times \frac{4}{47} \times \frac{3}{46} \times \frac{2}{45} \times \frac{1}{44}.$$

Our next question is a practical probability problem of an entirely different kind. I understand that this question was put to medical students in America, and the response was somewhat alarming.

We have a test for a particular disease that will certainly yield a positive result if the patient has the disease but also has a 5% chance of coming out positive if the disease is absent. It is known that one person in 1000 of the population has the disease in question. The problem is as follows.

2. A randomly chosen member of the population tests positive. What is the probability that this person has the disease?

Apparently many students gave the answer 0.95, the bland justification being that the test is 95% accurate. This really will not do. Their answer takes no account of the prevalence of the disease in the population and clearly that will affect the answer. For example, if the disease were smallpox, which has now been entirely eradicated, the answer would have to be 0—there would be no chance of a patient having smallpox *even if she tested positive*. We can see therefore that, if the disease is very rare, the chance of a positive test being a false positive would still be very high; the rarer it is, the greater the chance of a false positive. What then is the answer to our question?

The chances that a random person has the disease is one in 1000, that is 0.001. However, here we know more—we have a randomly chosen person of a special type—the type that tests positive: let us call such persons 'positive people'. The question is: what proportion of the positive people are actually diseased?

Consider a typical section of 1000 of the whole population (Figure 1). On average there will be one person with the disease, and a further 5% of the remainder, which to the nearest whole number is 50, who will yield false positives. What we have then is a randomly chosen member of the positive people, and we can now see that there is about 1 chance in 51 that the person suffers from the disease. It follows that the correct answer will be a little less than 2%—and not 95%!

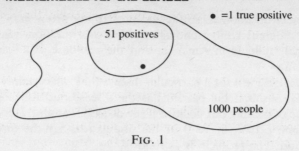

FIG. 1

Probability problems are often tricky, especially questions such as this one that involve *conditional probabilities* where you are asked for the likelihood of one event given that another has occurred. (In this example you want to know the probability that a person is diseased *given* that she tests positive.) Such problems can be quite treacherous—it is one thing not to be able to do a problem; it is quite another to imagine that you can do it and be led to an utterly wrong conclusion. This example shows how easily even intelligent and educated individuals can be fooled. It is worth having some people about who do understand mathematics.

The following diabolical conditional probability problem is old but manages to keep surfacing in different guises. It is sometimes known as the Monty Hall Problem, and a popular version of it goes as follows.

A contestant in a television game show is shown three numbered doors. Behind one lies the Big Prize while behind the others lie a goat. (Don't ask me why a goat.) The player picks a door. The host of the show, who knows what lies behind each door, then opens one of the *other* doors and shows the contestant a goat. The player then has the choice of either sticking with his original door or switching to the other which has not been opened. The question is:

3. Should the contestant stick or switch in the Monty Hall Problem?

The answer is yes, he really should switch, for it doubles his chances of winning! Most people, if not everyone, find this runs counter to their expectation. Why should the remaining unopened door be any more likely to house the prize than the one the contestant picked in the first place? Here is why the switch strategy is better.

The contestant makes an initial choice, let us say Door 1. The chances of his initial choice being correct are $\frac{1}{3}$. Monty Hall can always show you a goat behind one of the other doors, so the chances of Door 1 being the right choice are still $\frac{1}{3}$ after he has done this. Since the prize is not behind the door that Monty has opened, the chances of it being behind the third door must be $\frac{2}{3}$.

The mechanism for all this is made clearer if we increase the number of doors from 3 to 100. Since this is a *thought experiment*, we can afford to increase the number to 14 million. There is only one prize—all else is goats. You choose Door 1—you are almost certainly wrong: you have one chance in 14,000,000 to be precise. Monty now shows you goats behind *all* doors except Door 1 and one other. If you have not won the Lottery in the first place (a practical certainty) and there is a goat behind Door 1 also, he can only do this by showing you every other goat, in which case the prize is certainly behind the remaining door. Obviously you should switch, because the switch can only be wrong in the unlikely event that you chose right in the first place.

The argument is no different in the case of three doors, just that the probabilities are less extreme. If you are still unconvinced, try an experiment along these lines with a friend using, say, ten matchboxes or the like. It will not take many repetitions for the force of the above reasoning to make itself felt. There is a little more to be said, however, as my explanation implicitly assumes that, if Monty has a choice of two goats that he can show you (which is the case if Door 1 is the prize door), he then chooses at random; if he operated in another way, and if the contestant found out about it, he might be able to devise a better strategy. For instance, if we got to know that Monty was lazy and always showed what was behind the higher numbered of the two remaining doors if he could, choosing the lower number only if he had to so as not to reveal the prize, this indeed would be a valuable piece of information—in the event that the contestant chose Door 1 and Monty then showed him a goat behind Door 2, the player would know, for certain, that the prize lay behind Door 3 and would snap it up.

An equivalent version of this problem concerns three prisoners, Smith, Jones, and yourself. All have been sentenced to be executed in the morning. Whimsically, the Czar has decided to reprieve one of you: he has made his decision as to which one. In fact, the guard on watch has been told who is to live and die but the Czar, wanting to keep the good news a surprise, has forbidden the guard to reveal the truth.

You hatch a cunning plan to improve your chances. You approach the guard and say to him that you know he cannot tell you if you are the one to be saved but at least one of your comrades in the neighbouring cells is bound to be executed, so it would do no harm if he could reveal the name of someone, other than yourself, who has not been granted mercy. The guard takes pity and agrees, saying but one word: 'Jones'. You now reassure yourself with the following false reasoning:

> 'So poor Jones is heading for the chop. Ah well, that means that I now have a 50–50 chance since the other to be executed is just as likely to be Smith as it is me.'

Somehow it seems you have increased your chances of survival from 1 in 3 to 1 in 2 and so you sleep easier.

Of course, you have done yourself no good at all. (If this strategy worked, what would happen if all three of you used it!) The guard has simply shown you a goat behind one of the other doors—there is still only one chance in three that what lies behind your door in the morning is the prize of clemency. However, for Jones and Smith, who happened to be eavesdropping on your conversation with the guard, things are very different. Poor Jones has been 'shown the goat'—he is a doomed man, no doubt about it. Smith, on the other hand, has the right to feel somewhat relieved. Since your chance of survival is still $\frac{1}{3}$, his really must have improved to the complementary probability of $\frac{2}{3}$. Your little plan has not benefited you but it has done Smith some good. None the less, both you and Smith must wait till dawn when all doors are flung open to know your true fate.

Another less dramatic version of a similar question has been popular with authors of riddle books and mathematics examiners for years. You have a red ball and two yellow balls, numbered 1 and 2, in a hat. Your friend picks two of the balls out of the hat simultaneously and at random. What are the chances that both are yellow?

Since you are picking two balls out of the three and all are equally likely to be selected, the chances of the red ball coming up are $\frac{2}{3}$: therefore the chances that both are yellow (that is to say, red is *not* picked) are $\frac{1}{3}$.

Now suppose that you just happen to see some yellow between your friend's fingers as she drew the balls out. Given that extra information, what now are the chances that both balls are yellow?

The answer is still $\frac{1}{3}$, of course, as you have no extra information at all—you already knew that at least one of the balls must be yellow so your little peek has left you none the wiser.

Finally, suppose that you happened to spy not just some yellow but the number 1 on the yellow ball that was showing between your friend's fingers. What is the probability of two yellows now?

This really does change things! You know one of the balls is Y_1, the other is equally likely to be Y_2 or the red ball. Therefore the probability that she had drawn two yellow balls has increased from $\frac{1}{3}$ to $\frac{1}{2}$.

Why should it matter whether you know that the yellow ball you spied was numbered 1 or 2? The answer is that it doesn't matter which number it was—what matters is *knowing* which number it was.

Picking up on the theme of executions that arose in our previous question, let us consider the following one.

4. What is the probability of the first player winning in a game of Russian Roulette?

In case you are unfamiliar with this deadly pastime, let me explain the rules. Two players take turns to discharge a pistol pointed at their own head. Only one of the six barrels of the revolver is loaded. Before firing, the player spins the barrel so that the position of the bullet in the barrel is unknown. The players continue until one succeeds in shooting himself, where-upon he is declared the winner by the remaining player.

This is not a 'fair' game as the player who goes first has a slight 'advantage', but the question is: what is the exact probability that the first player fires the live round? We shall see in the next chapter that there is a natural way of solving this problem using geometric series. It is possible, however, to glean the answer immediately by exploiting the near symmetry of the situation.

Let the first player be A, the second B, and let a and b denote the respective probabilities that A and B win the contest. Of course, since the gun will go off sooner or later, we know that $a + b = 1$—it is certain that one or other player will win. Now the first shot of the contest is either fatal or it is not. If it is, B has zero chance of winning. However, there is a $\frac{5}{6}$ chance that it is not, in which case the tables are turned in that, in effect, A and B have swapped places and now it is B who enjoys the 'advantage'. In other words, in the event that the first shot is a blank, the probability that B will yet go on to win is a, the original prob-ability of A winning. This gives us a very simple equation relating a and b:

$$b = \frac{5}{6}a,$$

and coupling this with the fact that $b = 1 - a$, we get:

$$1 - a = \frac{5}{6}a \Rightarrow 1 = \frac{11}{6}a \Rightarrow a = \frac{6}{11}.$$

And so the first player has about a 54.5% chance of winning.

Let us next look at a problem which combines chance with geometry.

5. If a coin rolls across a chessboard, what is the likelihood that it will settle covering a corner of a square?

What is meant by this question is, if we were to repeat this experiment many times, what is the long-term proportion, as a number between 0 and 1, of the occasions when the coin would settle over some corner? The answer of course depends on the size of the coin. We shall assume here what would normally be the case in practice: that the diameter of the coin does not exceed the length of the sides of the squares of the board. We shall look at what happens should this not be the case a little later.

Again, to solve this problem it needs to be viewed from a different angle. The key observation on this occasion is that the coin will cover a corner if and only if the distance of the coin's centre from some corner does not exceed the radius of the coin: see Figure 2(*a*). The centre of the coin will lie in some square and it is equally likely

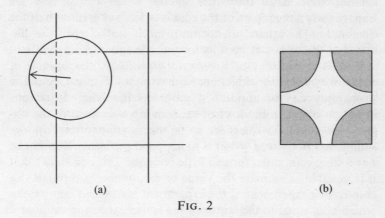

(a)   (b)

Fɪɢ. 2

to lie in one place as another. The area sufficiently close to a corner to cause the coin to cover that corner should the centre lie there is shaded in Figure 2(*b*). The probability that the coin will cover a corner then is just the ratio of the shaded area to the total area of the square. The shaded regions together comprise a circle equalling the area of the coin itself and so we can now formulate the answer: the probability that the coin covers a corner is

$$\frac{\text{(area of the coin)}}{\text{(area of each square)}}.$$

For a particular example, let us suppose that the diameter of the coin equals the side length of the square. Let us make this length 2 units so that the radius r of the coin is 1 unit. The area of the coin is $\pi r^2 = \pi$ and that of the square is $2^2 = 4$. Hence the answer in this instance is $\frac{\pi}{4} \approx 0.785$.

The problem is not too much harder if the coin is larger than the square. In principle we can solve it the same way, but now the four quarter-circles in the diagram above begin to overlap so that calculation of their total area is a little complicated, although still elementary enough. Mathematicians are sometimes guilty of ducking problems that become a little messy if they do not involve anything really new. It is perhaps worth noting though that we can answer the question as to how large the coin must be to ensure that it *must* cover a corner. This will be when the four quarter-circles cover the entire square, which we can now see happens when the radius of the coin is at least as great as half the diagonal of the square, or, to put it more neatly, when the diameter of the coin is at least as long as the square's diagonal.

This is a problem of *geometric probability*, the branch of mathematics dealing with chance behaviour of shapes. Geometric probability can be applied in problems involving deductions about an object from views of random cross-sections of the object—the object in question might be anything from an ore sample to brain tissue. What is more, good problems, like that of our rolling coin, often furnish little bonuses. This one shows that it is possible to estimate the value of the number π through this coin-rolling experiment: if the experiment is repeated many times with a coin equal to the width of the square, then the value of π will be approximated by four times the proportion of *successes* in the experiment, a success here being an outcome in which the coin covers a corner.

That the number π should emerge from this problem is none too surprising, as it does involve a circular object. However, π can also be estimated by the classical question of geometric probability, Buffon's Needle Problem, which seems to involve only straight lines.

The problem is this. You drop a needle on to the floorboards. What is the probability that the needle lands on a crack? Again, the answer depends on the length of your needle, and again it does involve π, and so an estimate of π can be found by noting the long-term proportion of cases where the needle does land on a crack. Without going through the calculation, I can give an explanation as to where the circular aspect of the problem lies that allows π to enter into the solution. Whether or not the needle hits a crack depends on two independent variables: the distance of the centre of the needle from the nearest crack, which is equally likely to be any value from 0 to half the width of the floorboards, and the *angle* the needle makes with a line through its centre and parallel to the line of the floorboards, which is equally likely to take on any value from 0 to 90°. It is the latter aspect of the calculation that introduces circular trigonometry into the problem and eventually leads to a solution involving π.

Finally, while on the subject of chessboards, consider the following question.

6. How many squares are on a chessboard?

This is not quite as trivial as it sounds, as, of course, we don't mean only the $8 \times 8 = 64$ unit squares but also the 2×2, 3×3, and all the larger squares as well. Again, like the coin-rolling problem, it is perhaps a little easier if we try to count another geometric feature which is in one-to-one correspondence with the one we are interested in. To be precise, it is easier, for example, to count all the 3×3 squares by counting their centres (Figure 3). A unit square itself forms the centre of a 3×3 square if and only

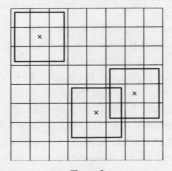

Fig. 3

if it does not lie on the edge of the board: these squares form a 6×6 mini-board within the main board so there are 36 of them. The centres of the 2×2 squares are all the corners of the squares of this central 6×6 mini-board—there are 7×7 such corners and so 49 2×2 squares. The sum of all the unit squares, 2×2 squares, and 3×3 squares is therefore $8^2 + 7^2 + 6^2$. You should not have much trouble convincing yourself that the total number of squares on the board is, perhaps not surprisingly, a sum of squares:

$$8^2 + 7^2 + 6^2 + 5^2 + 4^2 + 3^2 + 2^2 + 1^2 = 204.$$

This argument, of course, would similarly solve the problem for any square board of any dimensions. It would be nice, however, if we had a formula for the sum of squares as we do for the sum of the integers (Chapter 1, Problem 7). We shall find one in the next chapter.

Before introducing the next question, I shall indulge in a little preamble. If we have six children and five toys, we have a problem—at least two children will have to share. This is an instance of an extremely important principle in mathematics often known as the Pigeon Hole or Mail Slot Principle. This says that if we have n letters to be placed into m mail slots and $n > m$ (this means that n is greater than m), then at least one mail slot must contain two or more letters. To apply this to our party of children, we need to consider the toys as corresponding to the mail slots and the children to the letters—the difficulty is that $6 > 5$ and so at least one toy must be shared by two children (Figure 4).

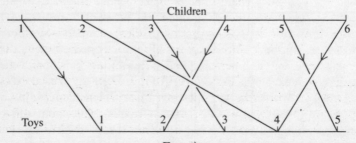

FIG. 4

This notion can be used to prove, beyond any doubt, things that at first sight look far from obvious. If a town has 400 people, then at least two of the residents must share the same birthday because there are more people than birthdays. There are at least two people in London with exactly the same number of hairs on their heads for the same reason: there are over seven million people in London but the number of hairs on anyone's head is no more than 250,000. (This is not obvious but is highly believable; if questioned we can afford to increase the allowed figure to several million and the Pigeon Hole Principle still yields the same conclusion.) In fact, we can say much more. There must be at least $6\frac{3}{4}$ million people in the capital for which there is at least one other person in the city with an equally haired head; the reason for this is that the number of people in London for which this is *false* cannot exceed 250,000. Already the principle is showing some of its subtlety. We shall use the idea behind it to tackle our next problem.

7. At a party must there always be some pair of people who have the same number of friends present?

Yes, there must. Let there be n people at the party. (Of course $n \geq 2$ as it takes at least two people to party.) The most number of friends a person can have at the party is $n - 1$; for example, the hostess might well be on good terms with all of her guests. The least number is 0. (This sounds sad, but is possible—there might be a gatecrasher, for instance.) Now suppose to the contrary that no two people at the party had the same number of friends. To each of the party-goers there is an associated number, we shall call it their *friend number*, which lies between 0 and $n - 1$ inclusive. We are supposing that all of these numbers are different from one another. This is not easy, but looks just possible: there are n different numbers to distribute among n people, which means that each of the n numbers $0, 1, 2, \ldots, n - 1$ is used exactly once. There is, however, one final twist which renders this impossible. Some person P scores 0 (no friends) while some other, Q say, scores $n - 1$. This means, however, that Q regards everyone else at the party, *including P*, as her friend. However, if P and Q

are friends, *P* cannot have a score of 0 after all. We therefore have reached the conclusion that the assumption of no two people at the party having the same number of friends leads to a contradiction, and so this assumption must be false. The only alternative is that there is a pair of people with equal numbers of friends at the gathering, and so it must be that way, for every party that ever there was, or ever will be, or ever could be.

We continue with a second party problem.

8.   At any party of six or more people, are there necessarily three mutual acquaintances or three mutual strangers?

The answer is yes, and the argument I shall give here to establish this is simple but quite delicate. The difficulty is that, given six people, there are many possible arrangements of acquaintance-ship that could arise between them. Our argument has to be able to cope with them all. If we go about it the wrong way we will be lost in a multitude of cases. Again, it is a matter of putting our finger on the key aspect of the problem.

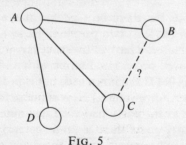

FIG. 5

Consider any six people at the party and focus on one of them, called A (Figure 5). Of the other five, either A knows at least three of them or, if not, there are at least three he does not know. (This is the only place where we use the fact that there are six people available.) Let us suppose for the moment that three of the people are known by A. Then either these three people are all strangers to one another, in which case we have found a required triangle of three mutually unacquainted people, or at least two of them, B and C say, do know each other. We then need only note that the three people A, B, and C form a triangle of mutual acquaintances. In the alternative case where there are three people who are not known to A, the argument is the same—you just need to run through it again swapping 'mutual acquaintances' for 'mutual strangers' throughout. We conclude that it is impossible to avoid a trio of mutual acquaintances or non-acquaintances when six or more individuals gather together.

We really do need at least six to make this argument work. To see this, imagine a party of five people sitting around a dinner table, and suppose that each person knows the two people sitting next to them, but not the other two guests. In this party there is no set of three who know one another, nor is there a trio who are totally unacquainted, as is easily seen by drawing a suitable picture.

Some of the problems we have looked at can easily be extended to larger numbers, but not so this one. To see what I mean, consider the same problem but this time ask yourself how large the party needs to be to ensure that there must be a group of *four* mutual acquaintances or four mutual strangers. You will not find it easy to generalize our approach at all. You may even begin to believe that there is no answer to the question—after all, it is conceivable that, however large the party, it might be possible to arrange things so that neither type of required foursome ever arises. That this is not so was proved by the English mathematician F. P. Ramsey in the 1930s. Ramsey's Theorem is a genuinely useful result in the mathematics of combinations which guarantees that, given any number m, at any sufficiently large gathering of people (the minimum size n depending on m) there is a clique of m people who are mutual acquaintances or mutual strangers. For instance, you need at least 18 people to *guarantee* such a clique of four persons—we say that the fourth Ramsey number is 18. No one knows the value of the fifth or any succeeding Ramsey number, but they do exist—Ramsey proved it.

Our ninth problem concerns a theme that we shall take up in the final chapter, that of *networks*. It is the classical problem of the

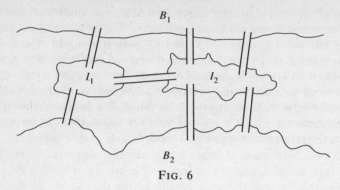

FIG. 6

Königsberg bridges. The old Prussian town of Königsberg lies on the banks of the Pregel River and is serviced by seven bridges which give access to the banks and to a pair of islands nestling in the river (see Figure 6). The question asked was:

9. Can one traverse all the bridges of Königsberg once and only once?

The frustrated citizens brought the problem to the attention of Euler, who explained why it could not be done. Although a simple problem, it was the first in the theory of networks and so required a fresh approach: until then such questions had not been seen for what they are—mathematical problems.

With respect to the bridge network, there are only four places a walker can be, as the lettering of Figure 6 indicates. Our first simplification in the way we look at the problem is to represent these four places (the two banks of the river and its two islands) as nodes or points on a diagram. For each bridge, we then, quite naturally, draw one line between the corresponding nodes, thus giving the diagram in Figure 7, which is simple and contains all the information relevant to the question in hand.

Suppose that there was some walk that traversed all the bridges exactly once. It would begin at a node and end at a node (perhaps the same one), but there would be at least two nodes that were at neither the beginning nor the end of the walk. Let X be such a node. Therefore we would visit X a number of times, and leave X *an equal number of times*. This would use up an even number of bridges in all—each time we arrived and left X we would use a pair of bridges that we would be forbidden to use again. It follows that X must have an even number of bridges servicing it. Unfortunately, however, this is true of none of the nodes in the picture: I_2 is connected to five bridges while the other nodes have three each. This shows that there is no walk with the properties we are looking for.

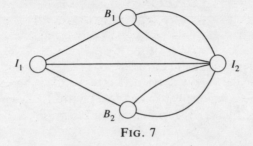

FIG. 7

This type of problem has since become popular as a puzzle: draw this figure without over-tracing a line (a bridge cannot be used twice) and without taking your pencil off the page (no jumps). This problem type we shall completely solve in Chapter 10, along with a collection of very different applications, some considerably more modern. In contrast, our next problem is very old indeed, being attributed to Heron of Alexandria about the year 75 of our era.

Mary lives at M and wants to visit her grandmother at G after taking a drink from the river, as depicted in Figure 8.

10. What is the shortest path that Mary may take for her journey?

Since the shortest distance between two points is a straight line, Mary's path will consist of two connected lines, the first from M

to some point P on the river and the second from P to G (Figure 8). The only question that remains is: how should she choose P?

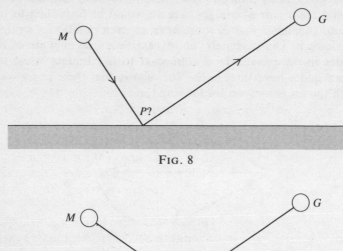

FIG. 8

FIG. 9

This can be a baffling question until we see that it is really one about reflections. Look at it this way. Suppose that Mary has a twin sister Maria who lives with her and who wants to visit her granny's twin who lives at G' directly opposite G on the other side of the river bank, exactly the same distance from the river bank as G. The sisters will travel together to some agreed point P on the river, enjoy their drink, and then part company for a while as Mary goes to G and Maria to G'. (Maria has the inconvenience of crossing the river, but that makes no difference to the problem's solution.)

Since G' is situated at the reflection of G in the line formed by

the river bank, the distances *PG* and *PG'* are identical as *PG'* is
merely the reflection of *PG*. It follows that we can minimize
Mary's walk by minimizing the length of Maria's journey, but
this is easy: for Maria to have the shortest possible trip, she
should travel in a straight line from *M* to *G'*. Thus we have
located the optimum point *P*: it is the intersection of the line of
the river bank with the line from *M* to *G'*, where *G'* is the reflection of the point *G* in the line of the river.

There is a truly substantive connection with this beautiful
problem and the behaviour of light. A beam of light sent from
M and striking a mirror placed at *P* with its face along the line of
the river will be reflected to *G* because, as we can see from Figure 9,
P is just so placed that the angle the line of the river makes with
MP equals that which it makes with *PG*. This shows the natural
efficiency of light as we see that the beam is taking the least time
possible to travel from *M* to *G* via the river bank.

This is an instance of Fermat's Principle of Least Time, which
also applies to light travelling through different refracting media
as represented in Figure 10. Here the light ray is travelling not on

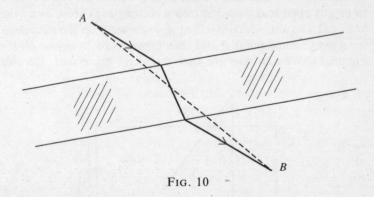

Fig. 10

the shortest path from *A* to *B* but on the one that requires the
least time: the straight line from *A* to *B* would involve the ray
having to travel further through the denser medium of glass
where its speed is less, and in consequence light travelling along
this path (if this were physically possible) would take longer to

reach B than that travelling on the path shown. From Fermat's Principle, one can derive Snell's Law concerning the ratio of the sines of the angles of incidence and refraction of a ray passing between two transparent media.

The idea behind this problem was to re-emerge in the nineteenth century in what would appear to be a totally unrelated context—that of finding the chances that the winning candidate in an election would lead all the way throughout the count. We shall see how to solve questions of this type using the Reflection Principle in Chapter 8.

We next have a problem that can be regarded in a similar light. An ant is on the outside of a cylindrical glass 4 inches high and 6 inches in circumference. Inside the glass, 1 inch from the top, is a drop of honey. Our ant is on the opposite side of the glass from the honey, 1 inch from the bottom.

11. How far must the ant walk to get to its honey?

The question is most easily dealt with if we imagine the cylinder to be cut open and flattened into a rectangle. (Throw away the bottom!) The ant, which starts at A, must walk up the outside of the glass to some point P and then down to its honey at H: see Figure 11. We can now see that this is no more than Heron's

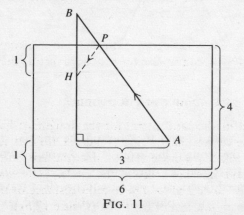

FIG. 11

Problem revisited, with the point A corresponding to the original M and H corresponding to G; and the question is to locate the unknown point P at the rim of the glass.

We again use the Reflection Principle: B corresponds to G' and so P is the point where the line from A to B meets the top edge of the glass. The length of the shortest path APH is then equal to AB, and by Pythagoras we find $AB^2 = 4^2 + 3^2$ so that AB is equal to 5 inches.

You should never be shy to question dubious arguments such as this one—although it is correct, it involves a leap of faith and this should at least be noted. We did not answer the question about the cylinder but about the rectangle that results from opening the cylinder up. Does this change the problem? Certainly if we tried to tackle similar questions about a sphere by 'flattening it out', the distortions resulting would lead to wrong answers. What is nice about a cylinder is that, in a sense that can be made mathematically precise, it is not really curved, and so flattening the curved surface of the cylinder causes no distortion. In particular, the length of any path on the cylinder does not change as it is flattened. Imagine yourself to be a piece of string on the surface of the cylinder which is held taut without being stretched. As the cylinder is opened out your shape will change from a curve to a straight line but your length remains the same—you will not be stretched, nor will you go floppy. This is why the two problems are equivalent and solving the second does solve the first.

If you are prepared to believe that we can do this kind of thing to cylinders, then we can solve more problems such as the following.

12. What is the volume of a doughnut?

The correct mathematical term for the doughnut shape is *torus*, known also as an anchor ring (Figure 12(a)). It is the shape formed by rotating a circle about a line (the axis) in the plane of the circle, the line not meeting the circle. Let r stand for the radius of the circle and let d denote the distance of the centre of the circle from the axis of the torus (Figure 12(b)).

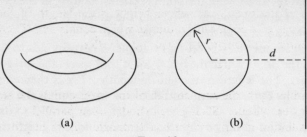

(a)   (b)

FIG. 12

The torus is one of the fundamental mathematical objects in the universe. This is not at all obvious and I say this partly by way of warning. Many books on mathematics, topology texts in particular, would have you agonize over whether certain things can or cannot be done on the surface of a torus, which may seem a perverse way to spend even the wettest of Sunday afternoons. However, these questions really do matter. I will not do much to prove this here; instead, let us now find the volume of our doughnut.

The idea is to slice the doughnut through a circular cross-section and straighten it out so that it becomes a cylinder with a half-cylinder truncated from either end (Figure 13). We can form a single cylinder from this object by mentally slicing off the half-cylinder from one end, turning it over, and replacing it at the other end so as to match the missing half-cylinder there. As you will remember, the volume of a cylinder is the area of its base times its height. In this case we have reconstructed our torus into

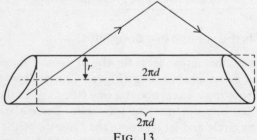

FIG. 13

a cylinder whose radius is r, the radius of the circular cross-section of the torus, and whose height is the length of the circumference of a circle with radius d, which equals $2\pi d$. Therefore the volume V of our torus is $(\pi r^2)(2\pi d)$, that is:

$$V = 2\pi^2 dr^2.$$

In a similar way, the surface area of the torus equals the surface area of our cylinder. Slicing the cylinder open parallel to its axis and flattening it out, we form a rectangle whose height is the same as that of our cylinder and whose width is the circumference of its base. The area of this rectangle, and hence the surface area S of the torus, is thus $(2\pi d)(2\pi r)$ which is

$$S = 4\pi^2 dr.$$

7

Series

Some examples of series

Some of the very simplest problems you first meet in mathematics involve spotting patterns in a sequence of numbers. This naturally leads to questions on series, the summing of number sequences, and one can quickly find oneself in very deep water, perhaps without realizing it. This book does not comprise a course in these matters, however, so in this chapter I am largely content with describing, rather than deriving, results on series. Where a series does none the less succumb to simple and short manipulations, a full explanation is offered.

Over the last few centuries a staggering amount of effort and diabolical ingenuity has been invested in problems that involve summing a series of numbers. Of course we can always add up any particular string of numbers—what I refer to here are problems of infinite series such as:

$$\frac{1}{2} + \frac{1}{4} + \frac{1}{8} + \ldots = 1, \tag{1}$$

or problems of finite series, such as the chess board problem of the previous chapter, where we ask for a formula in terms of n for the sum of the first n numbers of a certain type; in that case the problem required the summing of squares:

$$1^2 + 2^2 + 3^2 + \ldots + n^2 = \frac{1}{6}n(n+1)(2n+1). \tag{2}$$

While some series can be tamed only by using heavy mathematical machinery, others are accessible to simple algebra, including the two examples above, which we shall deal with a little later.

There is some explaining to do regarding infinite series since we cannot pretend to add the infinite series of numbers of (1) in the same way we can a finite series such as (2). Leaving this aside for the moment, I would like to begin with a list of examples to illustrate how series that are superficially similar can behave entirely differently. For the time being I will leave it to you, the reader, to spot the pattern of the terms in each of the following series—more detail will be revealed afterwards.

$$1 + \frac{1}{2} + \frac{1}{3} + \frac{1}{4} + \frac{1}{5} + \ldots = \infty \tag{3}$$

$$4 - \frac{4}{3} + \frac{4}{9} - \frac{4}{27} + \frac{4}{81} - \ldots = 3 \tag{4}$$

$$1 - \frac{1}{2} + \frac{1}{3} - \frac{1}{4} + \frac{1}{5} - \ldots = \ln 2 = 0.6931\ldots \tag{5}$$

$$1 - \frac{1}{3} + \frac{1}{5} - \frac{1}{7} + \frac{1}{9} - \ldots = \frac{\pi}{4} = 0.7854\ldots \tag{6}$$

$$1 + \frac{1}{4} + \frac{1}{9} + \frac{1}{16} + \frac{1}{25} + \ldots = \frac{\pi^2}{6} = 1.645\ldots \tag{7}$$

$$\frac{1}{2} + \frac{1}{6} + \frac{1}{12} + \frac{1}{20} + \frac{1}{30} + \ldots = 1 \tag{8}$$

$$1 + \frac{1}{8} + \frac{1}{27} + \frac{1}{64} + \frac{1}{125} + \ldots = ? \tag{9}$$

$$\frac{1}{2} - \frac{1}{6} + \frac{1}{24} - \frac{1}{120} + \frac{1}{720} - \ldots = \frac{1}{e} = 0.3679\ldots \tag{10}$$

$$\frac{1}{2} + \frac{2}{4} + \frac{3}{8} + \frac{4}{16} + \frac{5}{32} + \ldots = 2 \tag{11}$$

$$\frac{1}{2}+\frac{1}{3}+\frac{1}{5}+\frac{1}{7}+\frac{1}{11}+\frac{1}{13}+\frac{1}{17}+\ldots=\infty \qquad (12)$$

Series (3) The nth term here is $\frac{1}{n}$ and the series is known as the *harmonic series*. What do we mean in saying that the sum is infinite? Let us begin the explanation with a simpler example.

A series such as

$$1 + 1 + 1 + \ldots$$

obviously *diverges to infinity*, meaning that as we sum more and more terms of the series the sums we obtain increase beyond all bounds: in this case the sum of the first n terms is n. We have seen that this does not always happen and indeed, provided *the terms of the series approach zero*, sums taken from an infinite series of positive numbers *may* approach a limit: for instance look at our example (1)—as we sum more and more numbers from this series the resulting sums become ever closer to the limiting value of 1. We see then that the question of an *infinite* series converging to a limit arises only when the terms of the series approach zero. Now (3) satisfies this criterion: as n increases, the terms $\frac{1}{n}$ march steadily down to zero so it seems that there is a chance that the sum of the terms approaches a limiting value like we have seen in (1). However, this is not the case: the series diverges to infinity, meaning that, given any number, whether it be 10 or 10 million, that number will be exceeded if we sum enough terms from the series. This is not obvious, although I shall show that it is true a little later. The number of terms required to exceed 10 is well over 20,000 and the number needed to exceed 10 million doesn't bear thinking about.

What is the difference between the series (1) and (3) which may account for their contrasting behaviours? The important difference lies in the fact that the terms of the first series, $\frac{1}{2^n}$, converge to zero much more quickly than do the terms $\frac{1}{n}$. For large values of n both terms are of course very small, but the nth term of the latter is still many, many times larger than the nth term of the former. For example, take $n = 16 = 2^4$ (there being nothing special about 16 except that, being a power of 2, the following calculation is simple):

$$\frac{1}{16} \div \frac{1}{2^{16}} = \frac{2^{16}}{2^4} = 2^{12} = 4096,$$

and so, for $n = 16$, $\frac{1}{n}$ is thousands of times larger than $\frac{1}{2^n}$.

Series (4) This is an example of an (infinite) geometric series and is essentially similar to (1)—to pass from one term to the next you multiply by a fixed number, the *common ratio*, which in this case is $-\frac{1}{3}$. The nth term is $4(-\frac{1}{3})^{n-1}$; similarly, the general term of (1) is $(\frac{1}{2})^n$ while the common ratio is $\frac{1}{2}$. Geometric series are easy to handle and are important both in themselves and as tools for tackling more difficult series questions. I shall show how to sum geometric series later.

Series (5) This is the harmonic series with alternating sign. It is quite easy to convince oneself that this series does converge, meaning that successive sums become closer and closer to some *limit*. If we mark a few successive sums from the series on a

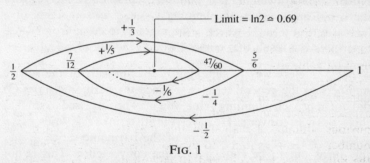

FIG. 1

number line, as in Figure 1, it becomes quite clear what is happening. Successive sums from the series jump either side of a limiting value, the jumps becoming ever smaller at every stage. This observation applies to any such series: if a series has alternating sign and the absolute size of each term is smaller than the term that preceded it, then the series converges. Indeed, we can say a little more: if we sum the first n terms of such a series, the difference between this sum and the sum of the entire series is no more than t_{n+1}, the next term of the series. In this example, for instance, the sum of the first five terms is:

$$1 - \frac{1}{2} + \frac{1}{3} - \frac{1}{4} + \frac{1}{5} = \frac{47}{60} = 0.78\dot{3},$$

which exceeds the limiting value by 0.0901... which is less than $\frac{1}{6}$, the next term in the sum. Again, examination of Figure 1 should help convince you of the truth of this observation: at any stage in the sum, the next term causes you to overshoot the limit, which shows that the sum of the first n terms of the series is closer to the limit than the size of t_{n+1}.

This does not, however, help us to find the *exact* value for the sum of the series (5), which is ln 2 (the 'ln' meaning the *natural logarithm*, which uses the special base of $e = 2.7183...$). This is clearly a harder nut to crack—after all, where is the logarithm going to come from? This result is not elementary and so lies outside the scope of this book—a little calculus is needed to do the job.

Series (6) Again we have an alternating series of ever decreasing terms so that, as in the last example, the series converges to a limit; but, once again, the limiting value, $\frac{\pi}{4}$, is something of a 'rabbit out of a hat'—where is the π going to come from? Once again this is a result that comes through the use of calculus.

Series (7) Here we are summing the reciprocals of the square numbers so the general term is $\frac{1}{n^2}$. Since the harmonic series (3) did not approach a limiting value, you may be surprised to see that this series does converge. However, the nth term of this series, $\frac{1}{n^2}$, is n times *smaller* than the nth term of the harmonic series, $\frac{1}{n}$, and this turns out to be enough to force convergence. This is not obvious, however, and requires some work. In this instance the argument is not beyond us but involves a little trick that you will see later. The mysterious limiting value, $\frac{\pi^2}{6}$, will, however, remain out of our reach. The usual method for obtaining this result employs techniques from what is known as *Fourier series*, which are important in the study of waves and periodic motion.

Series (8) Did you find the pattern here? The nth term in this case is $\frac{1}{n(n+1)}$. This looks almost the same as (7) and indeed we shall show how we can use the convergence of (8) to verify the convergence of (7). Fortunately, there is a simple algebraic trick that shows the sum of (8) is indeed 1, as you will see later.

Series (9) This is an obstinate series indeed. It is simply the sum of reciprocals of cubes and, as such, looks very similar to (7). Certainly it is not hard to show that the series converges to a limit, and it is possible to calculate that limit to any number of decimal places. What we lack, however, is an expression for the sum in terms of other numbers after the fashion of examples (5), (6), and (7). Indeed, the character of the limit was unknown until recent years when the French mathematician Apery proved that it was irrational. As for the sum of reciprocals of fifth and higher odd powers, not even that much has been decided. In contrast, it has long been known that the sums of reciprocals of even powers of positive integers can be expressed as rational multiples of powers of π (the sum (7) being an example), which are therefore irrational.

Series (10) Here is another alternating series, the nth term of which is $\frac{1}{(n+1)!}$, multiplied by ± 1 according as n is odd or even. This series converges *very* rapidly—the difference between the sum of the first n terms and the limit is always less than the next term, which is $\frac{1}{(n+2)!}$. For example, compare the sum of the first eight terms, which equals 0.367888, and the eventual limiting value of 0.367879... The difference is a mere 0.000009.

Series such as this which converge rapidly can be very useful tools in practical calculations. This particular series turns out to be the solution to the following curious problem. Suppose there are n different letters and n envelopes. A careless clerk, thinking that the letters are merely identical circulars, places one letter in each envelope at random. What is the probability that *none* of the letters will match the address on the envelope in which it is posted?

This problem, due to Euler, can be solved using what is known as the *Inclusion–Exclusion Principle*. We will not go further into that here except to say that the answer is the sum of the first n terms of our series (10). This has a surprising consequence which takes some effort to reconcile with intuition. The later terms in the series are so small they are practically negligible. This means that, provided that n, the number of letters, is more than about 4, the answer is virtually always the same and is very close to the limit of $\frac{1}{e} = 0.3679$. In other words, if there were 100 letters there is more than a 36% chance that the poor clerk got them *all* wrong! He undoubtedly would think himself cursed with extreme bad luck to be wrong 100 times out of 100, but unfortunately the mathematics is against him.

Series (11) The nth term in this case is $\frac{n}{2^n}$. Again a little calculus allows you to evaluate this sum. Calculus is not required, however: the result can be obtained by rewriting (11) as a series of geometric series and summing them all. This is one instance where the elementary approach involves more work than that using a more sophisticated technique. In mathematics the word 'elementary' does not necessarily mean easy—it means only that the problem was solved without recourse to higher mathematics.

Series (12) Here is a different series type—the sum of the reciprocals of the prime numbers. Just because there are infinitely many primes, as we showed in Chapter 4, it does not follow that this series diverges in the manner of the harmonic series (3). After all, there are infinitely many powers of 2 but the sum of the reciprocals of powers of 2, as was stated in (1), has a limit of 1. Suffice it to say that the standard proof that the sum of the reciprocals of the primes diverges is quite short and elementary, although it involves some sharp observation. I shall not include it here.

Finite series

In Chapter 1, Question 7, we saw that

$$1 + 2 + \ldots + n = \frac{1}{2}n(n + 1). \tag{13}$$

Having done this, we can sum the terms of any *arithmetic* sequence, that is any sequence that has a starting number a and where the difference between successive terms is a fixed number d:

$$a, a + d, a + 2d, \ldots, a + (n - 1)d, \ldots$$

Since the first term is a, the second $a + d$, etc., the nth term, denoted by t_n, is obtained by adding d to a a total of $n - 1$ times, so that $t_n = a + (n - 1)d$. The series of positive integers is the arithmetic series in which $a = d = 1$. We wish to find the sum of the arithmetic *series*:

$$a + (a + d) + (a + 2d) + \ldots + a + (n - 1)d. \qquad (14)$$

We can do this given our knowledge of the sum (13): the general sum is basically the same with a change of scale (the difference between successive numbers is changed from 1 to d) and a shift (instead of starting at 1 we begin at an arbitrary number a). These simple algebraic alterations are easily coped with as follows.

Taking all the as to the front in expression (14), we obtain, since there are n of them in all:

$$na + d + 2d + \ldots + (n - 1)d.$$

Take the common factor of d outside of all the later terms:

$$na + d(1 + 2 + \ldots + n - 1).$$

The formula (13) gives the sum of the first n counting numbers: to obtain the sum of the first $n - 1$ numbers we simply replace n by $n - 1$ throughout to obtain:

$$na + d\left(\frac{1}{2}(n - 1)n\right).$$

That is to say:

$$a + (a + d) + \ldots + a + (n - 1)d = na + \frac{d}{2}n(n - 1).$$

We can apply this to any arithmetic series we choose. For example, the sum of the first n odd integers, which is an arithmetic series in which $a = 1$ and the common difference $d = 2$, is:

$$1 + 3 + 5 + \ldots + (2n - 1) = n \times 1 + \frac{2}{2}n(n - 1)$$
$$= n + n(n - 1)$$
$$= n + n^2 - n = n^2,$$

which is a result we had already seen geometrically in Question 6 of the first chapter.

There is little more to know about summing arithmetic series, although it is worth pointing out that there is no restriction on the numbers a or d: they can be positive, negative, or zero.

A common question in IQ-style tests is to write down the next three numbers in a sequence such as

$$4, 7, 12, 19, 28, 39, 52, \ldots$$

The thing to spot is that the difference between successive terms is increasing, in this case by 2 each time, i.e. by

$$3, 5, 7, 9, 11, 13, \ldots$$

so the answer would be $67, 84, 103$. The given sequence is not itself an arithmetic sequence, but the sequence of differences is arithmetic.

Indeed, the sequence of squares:

$$1, 4, 9, 16, 25, 36, \ldots$$

is also of this kind as the sequence of differences is, as we have already seen twice, the arithmetic sequence of odd numbers.

Can we sum the first n square numbers and so derive the formula given in (2) like we can arithmetic series? Not imme-diately. We need to return to the problem of summing the inte-gers once more and solve it quite another way.

Consider the following strange-looking sum:

$$(1^2 - 0^2) + (2^2 - 1^2) + (3^2 - 2^2) + \ldots + (n^2 - (n-1)^2).$$

It is easy to simplify—the sum *telescopes* down to one term as, with the exception of n^2, every positive power is cancelled by its negative in the succeeding bracket. Hence the sum simplifies to n^2.

There is something to be learnt, however, if we pretend for a moment that we have not noticed this. A typical term in this sum is:

$$(m+1)^2 - m^2 = m^2 + 2m + 1 - m^2 = 2m + 1,$$

where m ranges from 0 to $n - 1$. It follows that this sum can also be written as:

$$1 + 3 + 5 + \ldots + 2n - 1,$$

and the previous telescoping argument shows that to equal n^2. This is the third time we have proved this fact, although this proof might strike you as rather artificial. None the less, it will prove to be a new and useful technique as it lends itself to generalization in a way that the others do not.

We should be curious enough to ask what would happen should we replace the difference of two squares with the difference of cubes. Do we get anything new? In other words, let us look at:

$$(1^3 - 0^3) + (2^3 - 1^3) + (3^3 - 2^3) + \ldots + (n^3 - (n-1)^3).$$

Again the sum telescopes down to a single term, this time n^3. The general term is $(m + 1)^3 - m^3$. As was shown in Chapter 5, we can expand and simplify this:

$$(m + 1)^3 - m^3 = m^3 + 3m^2 + 3m + 1 - m^3 = 3m^2 + 3m + 1.$$

Summing the general term $3m^2 + 3m + 1$ from $m = 0$ to $m = n - 1$ in fact gives us the sum of three sums, two of which we already know about while the other is the sum of squares that we seek. It follows that, with a little algebra, we shall be able to recover an expression for the sum of squares:

$$3(0^2 + 1^2 + \ldots + (n-1)^2) + 3(0 + 1 + \ldots + n - 1)$$
$$+ (1 + 1 + \ldots + 1) = n^3.$$

Now the sum of the n ones is of course equal to n. The sum of the integers from 0 up to $n - 1$ we know is given by $\frac{1}{2}n(n-1)$. Thus we get:

$$3(1^2 + 2^2 + \ldots + (n-1)^2) = n^3 - n - \frac{3}{2}n(n-1). \tag{15}$$

All that remains is to tidy up: $n^3 - n = n(n^2 - 1)$, and using the difference of two squares expression from Chapter 5 we may write this as $n(n-1)(n+1)$. Hence the right-hand side becomes:

$$n(n-1)(n+1) - \frac{3}{2}n(n-1).$$

These terms have a common factor of $n(n-1)$ which we extract to give:

$$n(n-1)\left((n+1) - \frac{3}{2}\right) = n(n-1)\left(n - \frac{1}{2}\right).$$

A neater expression comes from taking a factor of $\frac{1}{2}$ out of the last bracket. This is achieved by brute force—we think of n as $\frac{1}{2}2n$ to do this, so that $n - \frac{1}{2} = \frac{1}{2}(2n-1)$. We finish by dividing both sides of the resulting equation by 3 to isolate our sum of squares:

$$1^2 + 2^2 + 3^2 + \ldots + (n-1)^2 = \frac{1}{6}n(n-1)(2n-1). \qquad (16)$$

This is the sum of the first $n-1$ squares; if we want the sum of the first n squares, replace n by $n+1$ throughout the formula (16) and, with a reordering of the two central terms in the product, we obtain:

$$1^2 + 2^2 + \ldots + n^2 = \frac{1}{6}n(n+1)(2n+1). \qquad (17)$$

We can now go on and use this recursive technique to obtain the sum of the first n cubes, fourth powers, and in general kth powers: for cubes you would base the telescoping sum on $(m+1)^4 - m^4$, for instance. It turns out that:

$$1^3 + 2^3 + \ldots + n^3 = \frac{n^2}{4}(n+1)^2$$

and

$$1^4 + 2^4 + \ldots + n^4 = \frac{n}{30}(n+1)(2n+1)(3n^2 + 3n - 1).$$

The sum of integers formula is a quadratic polynomial in n, the sum of squares involves a cubic polynomial, and it is not hard to convince oneself (and not too hard to prove rigorously) that the formula for sums of kth powers will be a polynomial in n

involving n^{k+1} as its highest power. The problem of finding sums of kth powers comes down to determining the coefficients of these special polynomials. These coefficients can be given in terms of what are known as *Bernoulli numbers*, which often arise in problems of this flavour.

Geometric series

The most important class of series is the *geometric series*. They arise constantly in applications for they feature in the study of compound interest (indeed, they pervade much of elementary economics) and in topics such as population growth.

Perhaps the earliest problem of this kind occurs in the story, I believe Persian in origin, of the man who invented the game of chess. The King was so pleased with the new pastime that the inventor was asked to name his reward. Modestly, it seemed, he asked only for some grain—1 grain of wheat for the first square of the chess board, 2 grains for the second, 4 for the third, 8 for the fourth, and so on. The King gladly granted the request, only to find that he had agreed to give the man more grain than there was in the world, as we shall see shortly.

As with arithmetic sequences, a *geometric* sequence begins with an arbitrary initial term a, but this time it is the *ratio* of successive terms, and not the difference between them, that is constant. This *common ratio* is denoted by r. Thus, the first n terms of a typical geometric sequence take the form:

$$a, ar, ar^2, \ldots, ar^{n-1}.$$

For example, if $a = 1$ and $r = 2$ we have a geometric sequence that we have met before:

$$1, 2, 4, 8, 16, \ldots, 2^{n-1}, \ldots$$

If we sum a geometric sequence, we have a *geometric series*:

$$a + ar + ar^2 + ar^3 + \ldots + ar^{n-1}. \tag{18}$$

We can find a *closed* expression for (18), by which we mean an expression that has a fixed number of terms independent of the value of n, by using a trick which relates (18) to a certain telescoping sum. We multiply the geometric series by $1 - r$:

$$(a + ar + ar^2 + \ldots + ar^{n-1})(1 - r).$$

Then we expand this using the Distributive Law and obtain:

$$a + ar + ar^2 + ar^3 + \ldots + ar^{n-1}$$
$$-ar - ar^2 - ar^3 - \ldots - ar^{n-1} - ar^n,$$

where we see that only the first and last terms survive the cancellation to yield:

$$a - ar^n = a(1 - r^n).$$

Dividing both sides by $1 - r$ gives the answer:

$$a + ar + ar^2 + \ldots + ar^{n-1} = \frac{a(1 - r^n)}{1 - r}. \qquad (19)$$

We can now, for example, check the formula for the sums of powers of 2 which arose in our very first problem of Chapter 1:

$$1 + 2 + 2^2 + \ldots + 2^n = 2^{n+1} - 1. \qquad (20)$$

The left-hand side is a geometric series with $a = 1$, $r = 2$: the number of terms in the series is, however, $n + 1$, so we need to make that adjustment when applying the formula (19):

$$\frac{1(1 - 2^{n+1})}{1 - 2} = \frac{1 - 2^{n+1}}{-1} = 2^{n+1} - 1.$$

This also solves the classical grain-on-the-chessboard problem. The reward is given by equation (20) with n taken to be 63, so that the King owed

$$2^{64} - 1 \approx 1.84 \times 10^{19}$$

grains of wheat! The monarch was obviously unacquainted with the speed at which geometric series are liable to increase. (However, despite being made to look foolish, we are assured that the King was magnanimous and took it all rather well.)

Returning to the mathematics, the story serves to underline the fact that, if the common ratio r exceeds 1, then the series grows in an unbounded way as n increases. This does *not* happen, however, if r lies strictly between -1 and 1, because for very large values of n the term r^n, instead of dominating proceedings as it would if r were large, shrinks towards 0. In this case the sum of the series approaches a limiting value as n increases, and since the term r^n does not contribute to this limit we find that:

$$a + ar + ar^2 + \ldots = \frac{a}{1-r} \quad \text{if} -1 < r < 1. \qquad (21)$$

This allows us to check the infinite series that arose in the clock problem of Chapter 1. In that example $a = 1$ and $r = \frac{1}{12}$ so that:

$$1 + \frac{1}{12} + \frac{1}{12^2} + \ldots = \frac{1}{1 - \frac{1}{12}} = \frac{1}{\left(\frac{11}{12}\right)} = \frac{12}{11}.$$

Similarly, the reader may check that $1 + \frac{1}{2} + \frac{1}{4} + \ldots = 2$.

But we have already passed into the realm of infinite series.

Infinite series

Infinite series can be easier to handle than finite ones. For example, take the infinite geometric series:

$$L = a + ar + ar^2 + ar^3 + \ldots$$

Given that the series does have a limiting value L, we can express it in terms of a and r easily. We merely observe that:

$$L = a + r(a + ar + ar^2 + ar^3 + \ldots) = a + rL.$$

Solving the equation $L = a + rL$ gives us:

$$L - rL = a \Rightarrow L(1 - r) = a$$
$$\Rightarrow L = \frac{a}{1-r},$$

as we found in the previous section where we carefully investigated what transpired as we passed from the finite case to the infinite one. As we saw there, an infinite geometric series converges provided the common ratio r lies between -1 and 1, for this guarantees that the term r^n approaches 0 as n becomes large. For example, the series given in (4):

$$4 - \frac{4}{3} + \frac{4}{9} + \cdots$$

is a geometric series with $a = 4$ and $r = -\frac{1}{3}$, so that its sum is

$$\frac{a}{1-r} = \frac{4}{1+\frac{1}{3}} = 4 \times \frac{3}{4} = 3.$$

This is also an opportunity to revisit the Russian Roulette Problem (Question 4 of Chapter 6). Recall that there is one chance in six of the gun firing on each player's turn. The probability that the gun fails to fire for *both* players in any particular round is therefore:

$$\frac{5}{6} \times \frac{5}{6} = \frac{25}{36}.$$

The probability that player A wins on his $(n+1)$th shot after enduring n unsuccessful rounds is therefore:

$$\frac{1}{6} \times \left(\frac{25}{36}\right)^n, \quad n = 0, 1, 2, \ldots$$

The probability p that A wins is the sum of these individual probabilities over all possible values of n:

$$p = \frac{1}{6} + \frac{1}{6} \times \frac{25}{36} + \frac{1}{6} \times \left(\frac{25}{36}\right)^2 + \cdots$$

This is an infinite geometric series with the first term $a = \frac{1}{6}$ and $r = \frac{25}{36}$. Therefore we obtain:

$$p = \frac{a}{1-r} = \frac{1}{6} \div \left(1 - \frac{25}{36}\right) = \frac{1}{6} \div \frac{11}{36} = \frac{1}{6} \times \frac{36}{11} = \frac{6}{11},$$

which confirms our solution of the previous chapter. This

approach required a little more work but did reveal some extra information along the way.

Let us look at some infinite series that are not geometric. As we mentioned earlier, the harmonic series, the sum of the reciprocals of the positive integers, diverges; that is,

$$1 + \frac{1}{2} + \frac{1}{3} + \frac{1}{4} + \cdots$$

grows beyond all bounds (although extremely slowly). The terms of the series are small, but not as small as those in the previous example, for instance, and that causes it to behave quite differently. It is in fact easy enough to show that the harmonic series diverges. The standard argument involves grouping terms and comparing:

$$1 + \frac{1}{2} + \left(\frac{1}{3} + \frac{1}{4}\right) + \left(\frac{1}{5} + \frac{1}{6} + \frac{1}{7} + \frac{1}{8}\right) + \cdots$$
$$> 1 + \frac{1}{2} + \left(\frac{1}{4} + \frac{1}{4}\right) + \left(\frac{1}{8} + \frac{1}{8} + \frac{1}{8} + \frac{1}{8}\right) + \cdots$$
$$= 1 + \frac{1}{2} + \frac{1}{2} + \frac{1}{2} + \cdots$$

The reason why we bracket the terms in this manner is to create groups of terms each of whose sum exceeds $\frac{1}{2}$, and so the sum of the series increases beyond all bounds. Admittedly, we need to double the number of terms we take as we pass from one group to the next, but since the series is infinite this causes no difficulty. It follows that the sum has no limiting value.

Curiously, there is a simple and usable formula which allows us to calculate the sum of any number of terms from the harmonic series to a high degree of accuracy:

$$1 + \frac{1}{2} + \frac{1}{3} + \cdots + \frac{1}{n} \approx \frac{1}{2n} + \ln n + \gamma. \tag{22}$$

Once again, the expression $\ln n$ stands for the logarithm of n to the base e, but what is the mystery number γ? I'm afraid no one knows much about it. This number, known as the Euler–Mascheroni constant, certainly exists, by which we mean that the approximation (22) becomes ever more exact with increasing values of n. The constant γ itself can be calculated to any number of decimal places: to four places it equals 0.5772, so for instance:

$$1 + \frac{1}{2} + \frac{1}{3} + \ldots + \frac{1}{100} \approx \frac{1}{200} + \ln 100 + \gamma = 5.187.$$

However, even the basic question as to whether or not γ is rational has yet to be answered. Unlike other natural constants such as e and π, the number γ does not arise elsewhere in mathematics, making it difficult to come to grips with. Often new mathematical results are derived through being able to look at one thing in two different ways: the combined view from the two angles is often revealing. We still lack a second telling angle from which to attack γ.

We can use our knowledge of the divergence of the harmonic series to extend our result on Egyptian fractions of Chapter 5. Recall that we saw that any *proper* rational, $\frac{m}{n}$, could be written as the sum of the reciprocals of distinct positive numbers. We can now remove the restriction that $m < n$.

Let us consider $k \geq n$ so that $\frac{k}{n}$ is an improper fraction which we can write as a mixed number, $a + \frac{m}{n}$, where a is a positive integer and $\frac{m}{n}$ is a proper fraction. By the method of Chapter 5, we can write $\frac{m}{n}$ as the sum of m or fewer reciprocals of distinct positive numbers.

For example, suppose that our number is $\frac{16}{7} = 2 + \frac{2}{7}$ so that $a = 2$, $m = 2$, and $n = 7$. Our method gives:

$$\frac{2}{7} = \frac{1}{4} + \frac{1}{28}.$$

Having done this, we focus on a. Take the harmonic series and delete the reciprocals used in the sum for $\frac{m}{n}$. The remaining series still diverges, as removing a finite number of terms does not alter its divergent nature. By summing enough terms from the series, we may eventually exceed the given value a: we concentrate on the terms of the series that take us just past our target, a.

In our example, $a = 2$. Bearing in mind that we are forbidden to use $\frac{1}{4}$ and $\frac{1}{28}$ again, we drop them from the series and begin summing. We find that:

$$1 + \frac{1}{2} + \frac{1}{3} = 1\frac{5}{6} < 2 < 1 + \frac{1}{2} + \frac{1}{3} + \frac{1}{5} = 2\frac{1}{30}.$$

Thus, we see that:

$$2 = 1 + \frac{1}{2} + \frac{1}{3} + \frac{1}{6}.$$

If the fraction $\frac{1}{6}$ were not a unit fraction, we would now use the method of Chapter 5 to write it as a sum of distinct unit fractions. However, our example is completed, for, combining our decomposition of 2 with that for $\frac{1}{4}$, we arrive at:

$$\frac{16}{7} = 1 + \frac{1}{2} + \frac{1}{3} + \frac{1}{4} + \frac{1}{6} + \frac{1}{28}.$$

Thus we see there are three stages in the decomposition process: first decompose $\frac{m}{n}$; then decompose as large a part of a as possible, taking care not to repeat unit fractions; and finally deompose the remaining proper fractional part of a. We can always ensure that this final stage does not involve the use of any unit fractions used in the first two stages by using only those terms from the harmonic series that are sufficiently far along the series. For instance, if the unit fraction $\frac{1}{28}$ happened to arise again at the final stage, we could, in principle, decompose a again using only unit fractions whose denominator exceeded 28. The fact that the harmonic series diverges allows us to start as far along the series as we need. The number of terms needed for the decomposition might be very large, but a suitable decomposition could always be found.

Next we return, as promised, to the sum of reciprocals of squares. By way of contrast to the harmonic series, we shall show that the series (7),

$$1 + \frac{1}{4} + \frac{1}{9} + \frac{1}{16} + \ldots + \frac{1}{n^2} + \ldots,$$

converges. We shall first tackle the series in (8):

$$\frac{1}{1 \times 2} + \frac{1}{2 \times 3} + \frac{1}{3 \times 4} + \frac{1}{4 \times 5} + \ldots + \frac{1}{n(n+1)} + \ldots$$

The reason this series is more amenable to manipulation is that the general term, $\frac{1}{n(n+1)}$, can be written as $\frac{1}{n} - \frac{1}{n+1}$, a fact that can be verified by recombining over a common denominator:

$$\frac{1}{n} - \frac{1}{n+1} = \frac{(n+1) - n}{n(n+1)} = \frac{1}{n(n+1)}.$$

This allows us to write our series in a telescoping form:

$$\left(1 - \frac{1}{2}\right) + \left(\frac{1}{2} - \frac{1}{3}\right) + \left(\frac{1}{3} - \frac{1}{4}\right) + \left(\frac{1}{4} - \frac{1}{5}\right) + \dots$$

If we take the sum of the first n bracketed terms of this series, all terms are cancelled except the initial 1 and the final term $-\frac{1}{n+1}$; for example, the sum of the first four terms is plainly $1 - \frac{1}{5}$. In other words, the sum of the first n terms of this series is:

$$\frac{1}{1 \times 2} + \frac{1}{2 \times 3} + \dots + \frac{1}{n(n+1)} = 1 - \frac{1}{n+1}.$$

We conclude that this is a convergent series: the sum grows as we take more terms but never exceeds 1. In fact, since $\frac{1}{n+1}$ tends to 0 as n increases, it follows that the limiting value of these sums, which is what is meant by the sum of the infinite series, exists and is equal to 1.

We can now show that the original series of the sum of reciprocals of squares also converges through use of a comparison argument. Compare the two series:

$$\frac{1}{2 \times 2} + \frac{1}{3 \times 3} + \frac{1}{4 \times 4} + \frac{1}{5 \times 5} \dots + \frac{1}{n^2} + \dots$$

and

$$\frac{1}{1 \times 2} + \frac{1}{2 \times 3} + \frac{1}{3 \times 4} + \frac{1}{4 \times 5} + \dots + \frac{1}{n(n+1)} + \dots$$

If we compare the denominators of the corresponding terms of these series, we see that each denominator of the first series is *greater* than its counterpart in the second series. This means that each term of the first series is actually smaller than its counterpart in the second series. It follows that if we add the first n terms of the first series the sum will be smaller than the sum of the first n terms of the second. We have just shown that the sum of the first n terms of the second series is always less than 1. It follows that the sum of the first n terms of the first series is also always less than 1. Therefore the first series converges to a limit less than 1, the limit of the second series. We conclude that the sum of the reciprocals of the squares does converge and that:

$$1 + \frac{1}{2^2} + \frac{1}{3^2} + \frac{1}{4^2} + \ldots < 2.$$

I am afraid that we are not in a position to conjure up the magical-looking limiting value of $\frac{\pi^2}{6}$, but the series has another remarkable property none the less: the number of terms one has to add in order to get to within $\frac{1}{n}$ of the limit is exactly n. For example, the sum of the first ten terms lies within 0.1 of the limit but the sum of the first nine terms does not. Curiously, this can be proved without knowing the value of the limit using little more than the techniques you have seen above.

Compound interest and very long products

If we can have infinite sums, why not infinite products? There are some very pretty infinite products involving the number π. Perhaps the prettiest of them all is the seventeenth-century formula of John Wallis:

$$\frac{\pi}{4} = \frac{2}{3} \cdot \frac{4}{3} \cdot \frac{4}{5} \cdot \frac{6}{5} \cdot \frac{6}{7} \cdot \frac{8}{7} \cdot \frac{8}{9} \ldots$$

Just as we can evaluate only finite sums, we can evaluate only finite products. To say that this infinite product equals $\frac{\pi}{4}$ means that, as we evaluate longer and longer products from this expression, the answers we obtain become ever closer to $\frac{\pi}{4}$ and eventually approach within any given tolerance of this limiting value. Recall we observed that, for an infinite sum to have any chance of converging, the individual terms must approach 0. Similarly, for an infinite product to converge, the individual terms must approach 1. This is the case with the Wallis product: if you look at any pair of terms involving the same numerator, you will see that they have the form:

$$\frac{2m}{2m-1} \cdot \frac{2m}{2m+1} = \frac{4m^2}{4m^2-1},$$

so that the product of each such pair is just a little greater than 1. (For example, for $m = 5$ we have $\frac{100}{99}$ as our multiplier.)

The Wallis product comes from some mathematical trickery involving areas under curves of powers of trigonometric functions. Another infinite product involving π was discovered in the late sixteenth century by Viète. It comes merely from approximating circles by polygons, as you may suspect by the form it takes:

$$\frac{2}{\pi} = \sqrt{\frac{1}{2}} \cdot \sqrt{\frac{1}{2}\left(1+\sqrt{\frac{1}{2}}\right)} \cdot \sqrt{\frac{1}{2}\left(1+\sqrt{\frac{1}{2}\left(1+\sqrt{\frac{1}{2}}\right)}\right)} \cdots$$

These pretty mathematical ornaments may seem somewhat inconsequential. A much more mundane situation leads to a classical question involving the limiting behaviour of a product. This is Bernoulli's Compound Interest Problem.

Suppose you invest one unit (a unit could be a pound, a dollar, or a thousand pounds or dollars) in a scheme that pays 100% interest annually. (The actual interest rate makes little difference to the nature of the problem: I have selected this lucrative value only for ease of calculation.) After one year you will have 2 units. You would be better off, however, in a scheme that paid 50% twice yearly, as you would earn interest on the interest in the second half of the year—every six months your capital would be multiplied by a factor of $1\frac{1}{2}$ or, to put it another way, at the end of the year your account would hold:

$$\left(1+\frac{1}{2}\right)^2 = 2.25 \text{ units,}$$

an effective annual interest rate of 125%. Better still would be an account that paid interest monthly—your savings would be multiplied by $1\frac{1}{12}$ each month, yielding:

$$\left(1 + \frac{1}{12}\right)^{12} \approx 2.613 \text{ units,}$$

an annual percentage rate of 161.3%.

The shorter the waiting interval for your next interest payment, the better for the investor, so that if your account accrued interest daily, you would be better off still. Indeed, the bank could offer interest payment every hour or even every second. Why not take it to the limit and offer an account that accrues interest *continuously*. Is this possible? Would the bank instantly go broke because it would owe an infinite amount of money?

The answer is no, this *could* be done, as the worth of the customer's account will always be limited no matter how small the interval between interest payments is made.

The general situation is this: you are paid n times per year and each time your account is multiplied by the factor $1 + \frac{1}{n}$, so at the end of the year the number of units you will own is:

$$P = \left(1 + \frac{1}{n}\right)^n.$$

We have just seen that the larger we make n, the larger P will become. None the less, no matter how large a value of n is chosen, the value of P will remain less than 3. To show this in detail, which we shall not do here, one expands the above product for P using the Binomial Theorem (Chapter 4) and then observes that the terms of the expansion are each less than the corresponding terms of a certain geometric series which sums to 3. A little more work then reveals that the limiting value of P as n becomes ever larger is the number $e = 2.71828...$, the base of the system of natural logarithms.

8

Chance and Games of Chance

Birthdays and amazing lucky winners

If your little daughter returns from school saying that two children in her class both had birthdays today and asks, 'Isn't that amazing'? then the mathematically correct answer to her question is, 'No, that is not amazing, it is to be expected every once in a while.' Although this is not the way you might advise a parent to respond, it will be interesting to see why it is so, for the correct answer is surprising.

If we have two people, what are the chances they were born on the same day of the week? The answer is $\frac{1}{7}$—the first person 'chooses' a particular day (the weekday of his birth) and there is then one chance in seven that the second person's 'choice' will match that day. Another way of looking at it is that the probability that they were born on *different* weekdays is $1 - \frac{1}{7} = \frac{6}{7}$.

Suppose now we have three people: what are the chances that they were all born on different weekdays? This is the same type of problem as our solution to the National Lottery in Chapter 6. By the reasoning given there, the answer will be

$$\frac{6}{7} \times \frac{5}{7} = \frac{30}{49} \approx 0.61.$$

To find the chances that four people will be born on different days of the week, we multiply this figure by $\frac{4}{7}$ to give approximately 0.35.

The opposite of all the days being different is that some of them, two or perhaps more than two, are the same. Subtracting the previous probabilities from 1, we see therefore that the chances of two or more people being born on the same day is about 0.39 when we have three people, and 0.65 when there are four. It takes four people before we may expect a better than 50–50 chance of a coincidence of this type. Since 4 is the least number that exceeds half of 7, you might say that the answer is one you could expect. Note in passing that if we had eight people the probability of at least one coincidence would be 1—it could not be avoided, there being more people than days of the week. This is an incidence of the Pigeon Hole Principle introduced in Chapter 6.

All this may seem quite unremarkable, but it serves to illustrate the method by which we can answer the original question, 'Isn't it amazing?' How likely is it, in a class of 30 children say, that two or more will share the same birthday? The previous calculation concerning days of the week might suggest that the answer is 'very unlikely', for if that were anything to go by it might lead us to guess that you would need a group of at least half the number of days of the year, that is a class of 183 pupils, before we would have a better than even chance of there being a birthday coincidence. However, that is just a guess, and although the type of problem is exactly the same, the numbers are different, and so we would be jumping to conclusions. Ignoring the slight complication caused by leap years, the probability that the 30 children have 30 *different* birthdays is a product of the 29 fractions:

$$\frac{364}{365} \times \frac{363}{365} \times \frac{362}{365} \times \ldots \times \frac{337}{365} \times \frac{336}{365}.$$

This number turns out to be quite small, less than 0.3, so that the chance of two or more of the children having the same birthday is better than 7 in 10. Indeed, even in a class of only 23 the probability of a birthday coincidence is better than 50–50. This never ceases to surprise people, and the witnessing of what seems to be some astounding coincidence often comes down to not appreciating that these 'birthday' probabilities are not at all what you might first guess them to be.

It is true that I have assumed that birthdays are equally likely to fall on any day of the year. I expect this assumption of uniformity is pretty sound, although maternity hospitals that keep figures as to their busiest times of year may know better. However, departure from uniformity will not weaken the argument as that serves only to *increase* the likelihood of coincidence. To take an extreme imaginary example—suppose that we asked the same question of a society where men and women were obliged to live entirely separate lives except for one month of the year—July say; the conception of children would then be restricted to that month so that practically all birthdays would be nine months later, in April. It would then be a near certainty that in a group of thirty individuals at least two would share the same (April) birthday.

Often we hear of stories where someone has apparently unbelievable luck and the unlikelihood of the particular event is underlined with the assurance that it was a 'million to one shot'. With millions of chances being taken every day, it is true that the occasional extremely unlikely event will crop up by chance. Sometimes, however, these seemingly unlikely events are not especially unlikely at all. The apparent discrepancy can come down to not distinguishing between an improbable event happening to someone and happening to some specified person. For example, there is nothing amazing about *someone* winning the lottery, it is only amazing when it happens to a special person nominated in advance—you, for instance. Failure to grasp this point in more complicated situations can lead to some very perplexing outcomes.

For example, suppose you work for the giant firm Galactic Travel which rewards its 100,000 employees with a monthly lottery in which 100 of them win a free holiday. A computer chooses a name at random from the payroll, then returns to the list and picks again, 100 times in all. Conceivably, it might pick your name twice, but what are the chances? Well, the probability of being picked at all is no better than 1 in 1000, so that the chances of being chosen twice in the one month must be about one in a million. This reasoning is correct, so it is with some consternation that you read in the *Galactic Monthly Magazine* about Lucky Harry who won not one but two holidays in this month's draw! A one in a million shot! That was bad enough but, one year later to the day, you again read in the *Monthly* of Smart

Sally, another double holiday winner—and there to taunt you are the beaming faces of Harry congratulating Sally on her good fortune along with a list of this month's winners which, of course, never includes you. You feel that the chances against all this happening must be astronomic and go away muttering about the whole thing being fixed.

It is true that the luck of Harry and Sally is a little surprising, but only a little. You have to ask yourself: what is the probability that the computer will draw 100 *different* names from the list? This comes down to the Birthday Problem again, this time with 100,000 'birthdays' and 100 'pupils'; for 100 choices are being made from a possible 100,000 whereas in the original Birthday Problem the pupils were, in effect, making 30 random choices from 365 possible birthdays. The probability turns out to be around $\frac{19}{20}$ which, although quite high, leaves a 1 in 20 chance of one or more people being multiple winners. This means that a Lucky Harry or Sally can expect to crop up, on average, about one month in 20—that two such happenings should occur only 12 months apart instead of the expected 20 is a little unlikely, but no more than that.

And how about the fact that you never win? Of course you don't—after all, there is only one chance in 1000 of winning. If you kept backing horses that were 1000 to 1 against it would hardly be a surprise that none ever came in.

If this kind of thing really does upset you, it would be best to stop reading the magazine. If you don't, you will be tortured by 'amazing lucky winner' stories all your life—one month twin sisters will win, the next, someone will win his third holiday for the year. Since there are 100 lucky winners each month, one or two of them are liable to be especially lucky in some infuriating way. You just have to reconcile yourself to the fact that it will (probably) never be you.

Samuel Pepys's problem

Samuel Pepys, the famous keeper of diaries, was an earnest gambler and once posed to Isaac Newton the following practical gambling problem. In a game of dice, one man has six dice and is required to score at least one ace (a 1 on the showing face), while the second has twelve dice and needs to score two aces or more. Which player has the advantage?

I have the impression that Newton thought the problem was somewhat beneath him but he nevertheless gave Pepys his answer. You might suspect that there is enough symmetry in the game for it to be a fair one with neither player having an advantage, but Pepys's experience may have led him to suspect otherwise and, if so, he would have been right. One player does enjoy a small but definite advantage. Let us see which it is.

What is the first player's chance of failure? He will fail if every one of his dice shows a number other than 1. The chances that any particular die does this are $\frac{5}{6}$. Since each die behaves independently of all others, the proportion of times when all six of them show above 1 are $(\frac{5}{6})^6 \approx 0.335$. It follows that the first player's chance of success, that is of throwing at least one ace, is the complementary probability to this:

$$1 - 0.335 = 0.665.$$

What of the second player? This is a little more complicated. The second player may fail in one of two ways. Either he may throw no aces at all, or he may throw only one ace. Since he has 12 dice, the probability that he throws no aces is $(\frac{5}{6})^{12}$. We now require the probability that just one die shows an ace. The probability that the first die shows an ace (for it does no harm to imagine the dice being rolled one at a time, even if they are rolled simultaneously) while all others do not is:

$$\frac{1}{6} \times \frac{5}{6} \times \frac{5}{6} \times \ldots \times \frac{5}{6},$$

where there are 12 fractions in all corresponding to the 12 dice. Similarly, the probability that the second die shows 1 while all the others do not is:

$$\frac{5}{6} \times \frac{1}{6} \times \frac{5}{6} \times \ldots \times \frac{5}{6}.$$

Again, there are 12 factors and indeed, except for the order in which they are written, they are the same factors and so the answer will be the same. Since there are 12 such possibilities corresponding to the 12 places in the order of the dice where the ace could appear, we see that the probability that exactly one ace appears is:

$$12 \times \frac{1}{6} \times \left(\frac{5}{6}\right)^{11}.$$

To find the probability of success for the second player, therefore, we must subtract from 1 the separate probabilities of these two ways of failing, i.e. by throwing either no aces or exactly one ace. We obtain:

$$1 - \left(\frac{5}{6}\right)^{12} - 12 \times \frac{1}{6} \left(\frac{5}{6}\right)^{11} \approx 0.619.$$

We thus have answered Samuel Pepys's question—the first player has about 5% more chance of success than does the man with the 12 dice.

Counting problems, elections, and reflections

At the simplest level, probability questions involve a finite number of possible outcomes, all of which are equally likely, and we ask what is the probability that a certain favourable event will occur. In general, the required ratio is:

$$p = \frac{\text{number of favourable cases}}{\text{total number of cases}}.$$

We see immediately that p will always lie between 0 and 1, the extreme value of 0 corresponding to an impossible event, while if all cases were favourable so that success is guaranteed then p would be 1.

Often probabilities are expressed as percentages—a 50% likelihood of course would mean a probability of $\frac{1}{2}$. A common inaccuracy in probability jargon is sometimes heard in a situation where the speaker is reluctantly conceding that some undesirable outcome might just be possible. The admission often takes the form, 'There is a finite probability of that eventuality.' Since all probabilities are finite, this says nothing. What is meant, of course, is that there is a small but positive probability of the event occurring.

Calculation of probability p comes down to the problem of counting the numbers of total cases and of favourable cases. Questions of this type are very varied and interesting and can often be approached from a number of angles. Multiple dice questions are among the easier types. For example, what are the chances of rolling doubles in a two-dice roll? The total number of cases is in this instance $6 \times 6 = 36$ as each die has six outcomes. The favourable cases are six in number corresponding to both dice showing 1, both showing 2, and so on. Hence the required probability is $\frac{6}{36} = \frac{1}{6}$. Once the underlying set of possibilities for the experiment (in this case the roll of the dice) can be regarded as a collection of equally likely outcomes, such questions as these become simple matters of counting. For instance, the chance of rolling a total of 7, is also $\frac{1}{6}$ as six of the possible 36 outcomes yield a total of 7, while only two yield a total of 11, which therefore has a probability of only $\frac{2}{36} = \frac{1}{18}$ of occurring.

With card games, it is not so easy to do the required counting by inspection. For instance, what are the chances of being dealt a flush in a poker hand? (A poker hand consists of 5 cards from a normal deck of 52 and a flush is a hand composed entirely of one suit.)

Here, a little knowledge of binomial coefficients, as explained in Chapter 4, can go a long way. A poker hand is a choice of 5 cards out of 52 so the total number of hands is $C(52, 5)$, and this is the denominator of our probability ratio. As for the numerator, which is the count of the number of flushes, first we ask how many flushes there are in one particular suit, and then we multiply by 4 in order to count the total number of flushes over all four suits. The number of flushes in hearts, say, is the number of ways of choosing five cards from that 13-card suit, which is $C(13, 5)$. We can thus write down an expression for our probability p:

$$p = \frac{4C(13, 5)}{C(52, 5)} = \frac{4 \cdot 13!}{5! \, 8!} \cdot \frac{5! \, 47!}{52!}.$$

There is certainly no need to work out the huge numbers 52! and the like. This expression cancels massively: the two 5! terms cancel one another out and the 47! term cancels all but five of the numbers in the 52! term, leaving us with:

$$p = \frac{4 \cdot 13 \cdot 12 \cdot 11 \cdot 10 \cdot 9}{52 \cdot 51 \cdot 50 \cdot 49 \cdot 48} = \frac{33}{16,660} \approx 0.00198.$$

The probability is just under 0.2%—a rare event but not an incredible one.

We have solved this problem by regarding the choice of a hand as a single choice of a set of 5 cards from 52. As we did for the Lottery Problem, we can also solve it dynamically in the manner that we used for the Birthday Problem: imagine picking your cards up one at a time as they are dealt. The first determines the suit of any flush you might get; the probability that the second card matches that suit is $\frac{12}{51}$ (12 cards remain of that suit out of 51 still in the deck), and so on, giving the product of four fractions:

$$\frac{12}{51} \times \frac{11}{50} \times \frac{10}{49} \times \frac{9}{48},$$

and the same answer as before.

Our next problem looks, and to some extent is, of a similar type to the previous one but its solution, surprisingly, also draws on the reflection ideas seen in the Heron Problem of Chapter 6 and the question of the ant walking around a glass. At first sight this may seem rather bizarre in view of the nature of the question.

The votes are counted in an election in which there are two candidates, A and B, with A the eventual winner. What is the probability that A trails B at some point during the count?

The answer does of course depend on how many votes each candidate received. To make it exciting and simple, let us suppose that A polled $n + 1$ votes to n votes cast for B.

The first idea is to draw a picture describing the 'path' of the count. We draw a pair of axes, and plot points (x, y), where the point (x, y) indicates that after x votes were counted the lead of A was y votes. The values for x then range from 0 to the total number of votes cast which is $2n + 1$ in our case, and the values of y vary but are whole numbers which could of course be negative if B leads the count at some point. Finally, we join the plotted points simply to make the diagram clearer. All possible counts can be so pictured and every such path begins at O, with coordinates $(0, 0)$, and ends at $Q = (2n + 1, 1)$ as after all the $2n + 1$ votes are counted we know that A will emerge the victor by a single vote.

For example, if A polled 4 votes to 3 for B, two possible counts would be as pictured in Figure 1. The graphical picture of the

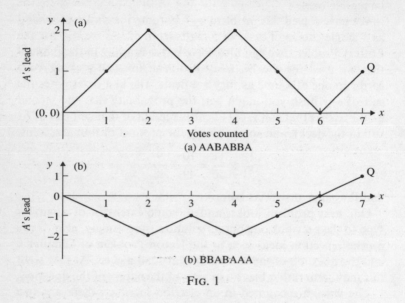

(a) AABABBA

(b) BBABAAA

Fig. 1

count is called its *lattice path* or simply its path (see Figure 2). The counts in which A trails at some point correspond to those lattice paths that touch or cross the line L which consists of all those points for which $y = -1$, for a y-value of -1 is indicative of B leading by one vote. Using a hash mark to indicate 'number of' we can write down an expression for the probability p that we seek:

$$p = \frac{\# \text{paths that touch or cross } L}{\# \text{paths}}.$$

It remains to find the two numbers in this ratio. The denom-

inator is easy enough. There are $2n + 1$ votes, of which n are for
B. A particular count is determined once we know where the
votes for B occur. The number of ways of choosing the n positions in the count where the votes for B turn up from the $2n + 1$
positions available is the binomial number $C(2n + 1, n)$.

Next we find the value of the numerator. Take a lattice path
that meets the line L and let P be the point on the path where it
first meets L. If we reflect the initial segment of the path from O
to P in the line while leaving the rest of the path untouched, the
result is a lattice path that begins at the point $(0, -2)$ and ends at
Q in Figure 2. It is equally the case that if we draw a lattice path
from $(0, -2)$ to Q it first meets the line L at some point P and if
we reflect that initial segment of the path in the line we obtain a
lattice path from O to Q which meets L.

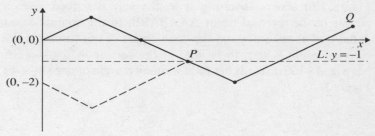

Fɪɢ. 2

The upshot of all this is that the number of lattice paths of the
type we are looking for exactly matches the number of lattice
paths from $(0, -2)$ to Q. The number of such paths is relatively
simple to count: since the path shows a net rise of 3 units from
beginning to end, there must be $n + 2$ places along the path where
it rises and only $n - 1$ places where it falls. The path is determined by the choice of the $n - 1$ places where the path falls out of
the $2n + 1$ places available, so that the total number of paths of
this type is given by the binomial coefficient $C(2n + 1, n - 1)$. We
can now find p:

$$p = \frac{C(2n + 1, n - 1)}{C(2n + 1, n)} = \frac{(2n + 1)!}{(n - 1)! (n + 2)!} \cdot \frac{n! (n + 1)!}{(2n + 1)!}.$$

Once again, most of the terms are kind enough to cancel themselves out, leaving behind the answer:

$$p = \frac{n}{n+2}.$$

For example, if A polled 99 votes to B's 98, that is to say if $n = 98$, we would get $p = 0.98$, indicating a 98% probability that B would lead A at some stage during the count, only to be disappointed in the finish.

It is also possible to say, without any further calculation, that there is a 98% chance that at some stage A would have led the count by more than one vote. This comes through consideration of the reversed counts as follows. Any lattice path can be regarded as representing the reversed count if we begin at Q instead of O and turn the diagram on its head. For example, consider Figure 1(b) above. Inverting it in the way described gives a picture of the reversed count AAABABB. In the original count A did trail at one stage and the counterpart of this feature in the reversed count is seen to be that at the corresponding stage A has a lead which exceeds her final winning margin of one vote (see Figure 3).

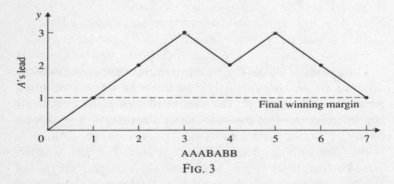

AAABABB

FIG. 3

This and other similar problems can be dealt with using this reflection technique. The original question of this type is called the Bertrand–Whitworth Ballot Problem and asks: if A and B poll a and b votes respectively with $a > b$, what is the probability that A leads the count throughout? This is a little more complicated than the problem we have solved but the solution proceeds along similar lines: the answer turns out to be $(a - b)/(a + b)$.

Ballot questions are a significant class of problems and arise in diverse settings such as particle physics and abstract algebra. It does not do to judge a mathematical idea by the context in which it is first illustrated, which may or may not be of intrinsic interest. A fresh idea that allows solution of a problem in an enlightening fashion merits respect.

Forcing a win at roulette

Sure-fire ways of winning at cards or at the roulette wheel are always in demand and there is one method that at first sight seems to work. The idea could be applied to any gambling game where unlimited stakes are allowed, but let us take roulette. The game is simply to place a stake on either black or red and if your colour comes up you have your stake returned plus the amount that you bet.

The strategy is simple—you back black until it wins. On the first spin you bet £1. If you lose, you bet £2 the next time. If you lose again next time round you bet £4 and so on. You continue doubling your stake stubbornly until eventually black does come up, whereupon you take your winnings and go home.

Does this make you a sure winner? Well, in a way the answer is 'yes'. If the wheel is fair, it really must show black sooner or later, say after n spins, where $n \geq 1$. How much have you lost on the previous $n - 1$ spins? Because of your double-or-nothing strategy, this gives a simple geometric series which we can sum using a formula from the previous chapter:

$$1 + 2 + 4 + \ldots + 2^{n-1} = 2^n - 1.$$

However, on the nth round you win £2^n, cancelling any accumulated losses and leaving you £1 better off than when you began. Not much, you may say, but you do win *for certain*, so if you are not satisfied you can play again and win another pound and go on until you have won as much money as you ever could want!

Would it really work in practice? The answer is that it almost certainly would. Having said that, I hasten to advise never to employ this strategy as you risk utter ruin for the sake of £1.

What can go wrong? The problem is that, although black must show eventually, there is always the chance that it won't show up until after you have lost all your money. To be sure, if you enter the casino with plenty of cash, let us say £10,000, the chance of this happening is very small. There would have to be 13 reds in a row before you were embarrassed by not being able to continue your strategy: 13 successive reds would see you accumulate a loss of $2^{13} - 1 = £8191$, and you would not have the cash on hand to double again.

You might scoff, saying this is not worth worrying about as the chance of 13 successive reds is 1 in a million. Such a run is unlikely, but not as unlikely as that: the exact figure is $\left(\frac{1}{2}\right)^{13} = \frac{1}{8192} \approx 0.00012$, something greater than 1 in 10,000. It is true none the less that you will almost certainly win your pound, but let's face it, if you have £10,000 to throw around, winning one more pound is no big deal and to do it you must risk much, much more.

The situation is the opposite of the Lottery. In the Lottery you take an enormous risk (for you almost certainly lose) on a small stake for the sake of having a microscopic chance of a huge win; in the above-described game of roulette you are taking a tiny risk on a huge stake for the sake of a having a near certain chance of a microscopic win. You are better off sticking with the Lottery.

World Series home game advantage

In American World Series baseball and basketball, the final two teams vying for the championship play a best-of-seven series of games which finishes once one team takes an unbeatable lead by winning four games. This year's finals are between, let us say, the Atlanta As and the Boston Bs. There is a perceived home team advantage, and for that reason great store is placed on the order in which games are allocated as home games for each team. There are a number of immeasurables here, but one popular belief is that there is an advantage in having your home games early in the series, in particular among the first four games, simply because you may never get to play the later ones; hence if your home games were scheduled for later in the series you may be denied the chance of exploiting the advantage they confer. This is a very reasonable and seductive argument, but I wish to show here that it is fallacious—there is no inherent advantage in the positioning of a team's home games within the seven-game series.

We can show this first through direct calculation. The point is made just as well through a simple example, so let us take it to be only a best-of-three series, and let us suppose that the As have only one home game to the Bs' two. Let us take the probability of the As winning on home turf to be p while the chances of an away win for the As is q—you can think of p as being greater than q, but the argument does not rely on this. The chances of a home loss for the As is the complementary probability of $1 - p$ and, similarly, their chance of losing away is $1 - q$. Writing h for home and a for away, consider two schedules for the As: haa and aah. According to the argument, the As should stand a greater chance of winning the title with the first schedule where they play their home game first, as in the second schedule they may never have the opportunity to play that home game. Let us work out the probabilities of the As winning the title under each regime.

Let us write W for a win for the As and L for a loss. Whatever the home-and-away arrangement, the As can win in three distinct ways: LWW, WLW, and WW—in this third case only two games are played as the third deciding match is unnecessary. Now let us operate under the haa schedule. The probability that the As will lose the first game is $1 - p$ as it would represent a home loss. The chances of an away win in the second game or the third game are q. Hence under the haa schedule the chance of the series going LWW for the As is the product of these three probabilities: $\Pr(LWW) = (1 - p)q^2$ (where Pr denotes 'the probability of'). Similarly, we can calculate that:

$$\Pr(WLW) = p(1 - q)q \quad \text{and} \quad \Pr(WW) = pq.$$

Therefore the probability that the As win the championship is:

$$(1 - p)q^2 + pq(1 - q) + pq = q^2 - 2pq^2 + 2pq. \quad (1)$$

Now let us operate under the aah timetable. Similar reasoning gives:

$$\Pr(LWW) + \Pr(WLW) + \Pr(WW) = (1 - q)qp + q(1 - q)p + q^2$$
$$= qp - q^2p + qp - q^2p + q^2$$
$$= q^2 - 2pq^2 + 2pq. \qquad (2)$$

The calculations represented by (1) and (2) give the same answer—there is no advantage for the As in playing the haa schedule as opposed to the aah alternative.

It might be said that our model is very simplistic. We assume constant probabilities for a win by the As depending only on whether they are home or away and independent of all other factors including previous results. This is unrealistic. To argue this way, however, is entirely to miss the point. If the *principle* that playing late home games was a disadvantage were true, it would apply to this model, and the disadvantage would surface in the above calculations. As it did not, the principle is not valid.

The above algebra shows that the probabilities of victory for the As is the same under both schedules, but it does little to explain why. It remains unclear as to why the initial argument that concluded that early home games should render an advantage is in general false. It helps to look at things in the following way. Imagine that the teams decide that they will play all the games of the series come what may. Since there is no longer any possibility of missing out on some of your scheduled home games, the original argument no longer applies and so there is no prima facie reason why early home games should confer an advantage. However, this imaginary change could not alter the likelihood of the As winning under any particular schedule as it takes effect only *after* the title has been decided. We are led to the conclusion that the apparent advantage of early home games was a mirage.

Match-my-best

This type of game is often played by two players at a basketball ring but it could be applied to any contest of skill. Players take turns to attempt some kind of trick shot of their own choice. If successful, the second player has to try to perform the same feat and, should he fail, the first player scores a point. The question arises: should you attempt easy tricks or hard tricks? Common sense tells us that a player would do best to attempt tricks that he finds relatively easy and, for whatever reason, his opponent finds difficult. I am sure this is the case, but there remains the question: all things being equal, what strategy confers the greatest likelihood of your scoring?

Let us look at the simplest situation in which you and your opponent are equally matched so that both of you have the same probability x of succeeding in any particular trick shot. You will score exactly when, on your turn to lead, you succeed but your opponent fails. Since the probability of failure is $1 - x$, the probability of your scoring with a shot of difficulty x is then $x(1 - x)$; for example, if you attempt a shot in which there is a one in three chance of success, the probability that you score is $\frac{1}{3} \cdot \frac{2}{3} = \frac{2}{9}$. What we need to do therefore is to find the value of x that maximizes the expression $x(1 - x) = x - x^2$.

This can be done using the algebraic technique of completing the square introduced in Chapter 5 to solve quadratic equations. The coefficient of x here is 1, so we add and subtract the square of $\frac{1}{2}$ to rewrite the expression as:

$$x - x^2 = -(x^2 - x) = -\left(x^2 - x + \frac{1}{4}\right) + \frac{1}{4}$$
$$= \frac{1}{4} - \left(x - \frac{1}{2}\right)^2.$$

Since a square is always non-negative, we maximize the expression through minimizing the subtracted square by putting $x - \frac{1}{2} = 0$; in other words, the best you can do is to take $x = \frac{1}{2}$, in which case your chance of scoring on your turn is $\frac{1}{4}$. For this game, mediocrity is the best strategy.

Games and game theory

There are not many of us, especially among people who like science, who have not watched *Star Trek* over the years. Each series has always included a character who behaves mechanically. In the original episodes it was Mr Spock, a man from the planet Vulcan who had banished all emotional behaviour from his psyche, while in the *New Generation* we have Data, a polite but ultimately emotion-free robot. Mr Spock assured us that he always acted *logically*. I cannot recall the character ever explaining what he meant by this, but we were left to conclude that he lived by some set of Principles and Values and always attempted to act in a way that was consistent with them. He would never act irrationally; that is to say he would not knowingly contradict his own principles, nor would he do something harmless but silly as he would have no impulse to act that way.

The humans in these series had to concede that both these characters were in general physically and mentally superior to themselves. There was one conceit, however, that the human characters treasured throughout. The humans would always insist that their ability to act illogically was sometimes to their advantage. There may be some truth in this, but most of the examples presented in the series were, in my opinion, entirely fallacious. The flaw lies in the assumption that unpredictability is akin to irrationality, which in many realistic situations is far from true.

In some games, notably poker, it is important that your play is, to some extent, unpredictable. This is not the same, however, as playing irrationally. The android Data is forever losing at poker, apparently because he cannot cope with the 'illogical' bluffing of his human opponents. This only means that he has been pro-grammed poorly. In a game such as poker it is important not to pass information about your hand to your opponent. To an extent, however, you are obliged to do this when you bet because, generally speaking, a player will be prepared to risk more when he holds a stronger hand than when he has a weaker one. If a player *never* bluffs his opponents will observe this and so be able to read the strength of his hand from the size of his stake, thus placing the 'logical' player at a disadvantage. There is nothing inherently illogical about bluffing in poker; it is in the nature of the game and forms a necessary part of a good strategy. There is no reason why a judicious element of bluffing should not be built into the playing strategy of a computer.

When a mathematician, economist, or military strategist speaks of an optimal strategy for a 'game', I believe many listeners would immediately assume that such a strategy involves adopting the same, supposedly best, response to any given scenario that may arise during the course of the game in question. In real-life games, and in real-life game theory, this is rarely the case—it is normally important that your opponent cannot predict your response with certainty. The element of surprise is in itself valuable and so should not be surrendered to your opponent without exacting a price.

This is clear in games much simpler than poker, such as the two-player game 'Paper, Stone, and Scissors'. Here players *simultaneously* show one of paper, stone, or scissors by means of a hand signal. Paper 'covers' stone which 'blunts' scissors which in turn 'cut' paper, and a player scores when his hand dominates that of his opponent. It is clear that you cannot afford to play predictably in this game; for instance, if your calls followed a fixed cycle, say, paper, stone, and then scissors in that order, time and again, your opponent would soon notice and adopt the cycle of calls scissors, paper, then stone, beating you every time. A measure of randomness must be present in any good strategy for 'Paper, Stone, and Scissors'. There is nothing illogical about this.

Some games, simpler than poker but more complex than Paper, Stone, and Scissors, can be analysed mathematically and best strategies actually calculated. An example, nicely explained in Bronowski's classic television series *The Ascent of Man*, is the game of Morra. In its simplest form each player shows one or two fingers and simultaneously guesses the number of fingers shown by her opponent, giving four options in all: (1, 1), (1, 2), (2, 1) and (2, 2), where by option (2, 1) we mean the player shows two fingers while guessing that her opponent shows one. If both players guess their opponent's hand correctly or if both guess incorrectly there is no score; a player scores only if she guesses her opponent's hand while her opponent's guess is wrong. In this case the player who got the call right wins the amount equal to the *total* number of fingers shown by both players.

The best strategy is to totally ignore options (1, 1) and (2, 2) and to use the (1, 2) and (2, 1) options, at random, but in the overall ratio of 7 to 5. How would you do this? You would need a *random number generator* (many calculators have one) that is arranged to generate random integers from 1 to 12. Ignore the behaviour of your opponent and play (1, 2) if the random number generated lies in the range 1–7 and play the alternative (2, 1) call if a number in the 8–12 range turns up. This does not of course guarantee that you will win a particular game—that will depend on your luck. If, however, you stick with this strategy, in the long run, as you play many many games, it is unbeatable. The best your opponent can expect to do in the long term is break even.

There is considerable calculation involved in deriving this strategy, and substantial mathematics in proving that it is the best. It is, however, easy enough to check that it is effective whatever the strategy adopted by your opponent. If the other player always sticks to the pair of strategies (1, 2) and (2, 1) like you do, then neither of you will ever win anything as you will either both call correctly (this happens if the two of you make *different* calls), or you will both be wrong (if you both choose the same call). Scoring will occur only when your opponent calls (1, 1) or (2, 2). Suppose she calls (1, 1): then 7 times out of 12 you will call (1, 2) and will lose 2 points in consequence; however, 5 times out 12 you will call (2, 1) and will win 3 points. On average, therefore, for every 12 times the other player calls (1, 1), you will have a net gain of:

$$5 \times 3 - 7 \times 2 = 15 - 14 = 1 \text{ point.}$$

A similar calculation shows that you will win when she goes for the (2, 2) option: she will win 4 points on the 5 out of 12 occasions when you play (2, 1) but will lose 3 points on the 7 out of 12 times that you play (1, 2), so that your long-term average gain on this type of play will be:

$$7 \times 3 - 5 \times 4 = 21 - 20 = 1 \text{ point.}$$

The calculations clearly show that the balance is in favour of your strategy, but this would not be at all obvious to an uninformed player even after considerable experience in playing the game—most gamblers would surely conclude that the (1, 1) and (2, 2) plays should be employed occasionally, but they would be wrong.

And so we see that unpredictability can be rational. There are some situations, however, where an irrational opponent is a stronger adversary than a rational one. Take a hostage crisis. Imagine you are a police negotiator trying to arrest Mr Spock, who is threatening to kill a hostage. You could argue:

> 'Spock, your only option is to surrender! If the hostage were killed you would be seized in any case and would suffer a more severe penalty in consequence. Your threat is therefore illogical. You will not execute it as you have no reason to carry it out.'

Logical Spock would be powerless to refute you and so you could arrest him at your leisure. If, on the other hand, you were dealing with a homicidal maniac, you would be in real difficulties. You could advance the same argument but would meet with the following refutation from your opponent:

> 'Ah, your argument is flawed. I am not a logical being but a maniac who requires no reasons! I still have the power to carry out my threat as, unlike Spock, I am immune to your reasoning.' (Or words to this effect.)

The maniac's argument is watertight. If you attempt to arrest him you are truly risking the life of the hostage. The difficulty lies in that the maniac, being human, can, unlike Spock, harbour simultaneously contradictory desires. On the one hand, he may not wish to suffer a more severe punishment any more than Spock, but on the other, he may need to vent his powerful feelings of anger and revenge. Who knows? No one can predict which of his impulses will dominate in the crucial moment of conflict—his actions may be totally unpredictable, even to himself. His more complex and fickle nature presents grave difficulties for the negotiator. He is certainly a more formidable opponent than Mr Spock in the Hostage Game.

Being the strongest player does not necessarily mean that you are in the most favoured position to win a game. This may sound like a contradiction, but this type of paradox often arises in games with more than two players, such as diplomacy involving several nations. In these situations it is good to be strong without appearing too threatening. If you appear to pose an intolerable threat to the other players, they may form an alliance and eliminate you entirely.

An example of this kind of game is the multi-player shoot-out, in which players have varying degrees of marksmanship which are *known* to all the players. The poor shots can turn out to be the stronger players as the top shots have no choice but to turn their guns on one another, invariably leading to total or near total annihilation of the better marksmen and leaving the poorer shots to fight it out. Indeed, in a three-way duel the poorer shot may, under certain probability regimes, be better off firing into the air.

Perhaps the most important example type in game theory goes by the name of the *Prisoners' Dilemma*. It is sometimes called a paradox because it shows that it is possible for a policy of self-interest to be inferior to a policy of cooperation for *every* individual in the community. It is often described in terms of two prisoners who face certain consequences in confessing or not confessing to a crime, the outcome for each being dependent not only on their own decision but also on that of their fellow prisoner.

The choices that the prisoners face are similar to those in a game whose outcomes are shown in Figure 4. Player A and

	B: 1	B: 2
A: 1	(5, 5)	(0, 20)
A: 2	(20, 0)	(1, 1)

FIG. 4

player B have two options: they can each write down the number 1 or the number 2. They act simultaneously. The payoffs for each player are shown in the table. For instance, if player A writes 1 while B writes 2, then A gets nothing while B earns £20. The game is to be played only once. What should they do?

Let us consider player A's position. He cannot control player B although they are free to talk things over beforehand and perhaps agree a 'deal'. However, when it comes to the crucial moment, each can choose whatever number he likes. Mr A wants the best deal for himself and reasons as follows. If B plays 1, then I win £5 if I play 1 but £20 if I play 2, and so in this case it is better for me to play 2. The alternative is that B will play 2, in which case I win nothing if I play 1 but £1 if I play 2. It follows that, *irrespective of what B does*, I will be better off to play 2, so that is what I will do.

The game is entirely symmetric of course, so B reasons the same way and consequently they both play 2 and win just £1 each. Silly boys! If they had just co-operated and both played 1, they would have got £5 each, but they just couldn't trust each other. But why should they? After all, the logic of the previous paragraph is irrefutable. Each player should try to persuade the other to play 1, but if they follow self-interest they should still cast their votes for 2. I am afraid that is just the way the game is—that is the Prisoners' Dilemma.

It is a different matter if the game is to be played many times, for then it really does make sense to cooperate. The players should play opposite strategies by turn and so alternate in collecting £20 rewards. This way each player earns an average of £10 per game which is better than the £5 each on offer if the (1, 1) 'cooperative' strategy is employed. However, when the number of games remaining begins to dwindle, the logic of *immediate* self-interest will again come to the fore and the players may once more be induced to cut each other's throats—they are liable to fall back into the 'logical' (2, 2) trap.

9

The Golden Ratio

In Chapter 2 it was said that although $\sqrt{2}$ is not a recurring decimal it does have a recurring expansion of another kind. I will now explain how that can be realized.

We begin by writing $\sqrt{2} = 1 + (\sqrt{2} - 1)$. We then think of $\sqrt{2} - 1$ as the reciprocal of its reciprocal. This may sound perverse, but have patience:

$$\sqrt{2} - 1 = \frac{1}{1/(\sqrt{2} - 1)}.$$

There is now a standard piece of algebra that allows you to do something interesting. That is the process of *rationalizing the denominator* as applied to the fraction $\frac{1}{(\sqrt{2}-1)}$. We multiply top and bottom by the *conjugate*, in this case $\sqrt{2} + 1$; for when expanded the denominator will be free of square roots as the change of sign will cause the cross-terms to vanish:

$$\frac{1}{\sqrt{2} - 1} = \frac{1}{\sqrt{2} - 1} \cdot \frac{\sqrt{2} + 1}{\sqrt{2} + 1} = \frac{\sqrt{2} + 1}{2 + \sqrt{2} - \sqrt{2} - 1} = 1 + \sqrt{2}.$$

In this case the new denominator has turned out to be 1. This gives us

$$\sqrt{2} = 1 + (\sqrt{2} - 1) = 1 + \frac{1}{1 + \sqrt{2}}.$$

We can now replace the new occurrence of $\sqrt{2}$ by $1 + (\sqrt{2} - 1)$ and then repeat the process indefinitely, yielding:

$$\sqrt{2} = 1 + \cfrac{1}{1 + \sqrt{2}} = 1 + \cfrac{1}{1 + 1 + (\sqrt{2} - 1)} = 1 + \cfrac{1}{2 + \frac{1}{1 + \sqrt{2}}}$$

$$= 1 + \cfrac{1}{2 + \frac{1}{2 + (\sqrt{2} - 1)}} = 1 + \cfrac{1}{2 + \frac{1}{2 + \frac{1}{1 + \sqrt{2}}}} = \ldots$$

We say that the *continued fraction* expansion of $\sqrt{2}$ is:

$$\sqrt{2} = 1 + \cfrac{1}{2 + \cfrac{1}{2 + \frac{1}{2 + \frac{1}{2}}}}.$$

We even use a notation reminiscent of recurring decimal notation to represent this: we write $\sqrt{2} = [1, \dot{2}]$. By truncating this representation after a particular number of divisions, we obtain a rational approximation of $\sqrt{2}$ (which to three decimal places is 1.414). Using just 1, 2, and 3 stages, we get the following fractions:

$$1 + \cfrac{1}{2 + 1} = \frac{4}{3} = 1.333\ldots, 1 + \cfrac{1}{2 + \frac{1}{2+1}} = \frac{10}{7} = 1.428\ldots,$$

$$1 + \cfrac{1}{2 + \cfrac{1}{2 + \frac{1}{2+1}}} = \frac{24}{17} = 1.412\ldots$$

What we have stumbled into here is actually another application of the Euclidean Algorithm of Chapter 4. To explain this, let us begin again with a rational number, $\frac{92}{73}$, and see how to construct its continued fraction expansion. First, the Euclidean Algorithm is applied to the pair of numbers 92 and 73, the result in this case being the following:

$$92 = 1 \times \underline{73} + \underline{19}$$
$$73 = 3 \times \underline{19} + \underline{16}$$
$$19 = 1 \times \underline{16} + \underline{3}$$
$$16 = 5 \times \underline{3} + \underline{1}.$$

We see that the highest common factor of 92 and 73 is 1,

indicating that the fraction is in lowest terms. We can now build a continued fraction expansion for $\frac{92}{73}$ as follows. Beginning with the first line of the algorithm, we have:

$$\frac{92}{73} = 1 + \frac{19}{73} = 1 + \frac{1}{\frac{73}{19}}. \tag{1}$$

From the second line we obtain:

$$\frac{73}{19} = 3 + \frac{16}{19}, \tag{2}$$

and substituting into the last equation we obtain:

$$\frac{92}{73} = 1 + \frac{1}{3 + \frac{16}{19}} = 1 + \frac{1}{3 + \frac{1}{\frac{19}{16}}}. \tag{3}$$

Using the third line, we then get:

$$\frac{19}{16} = 1 + \frac{3}{16},$$

and substitution gives:

$$\frac{92}{73} = 1 + \frac{1}{3 + \frac{1}{1 + \frac{3}{16}}},$$

and finally the fourth line gives $\frac{16}{3} = 5 + \frac{1}{3}$, whence

$$\frac{92}{73} = 1 + \frac{1}{3 + \frac{1}{1 + \frac{1}{5 + \frac{1}{3}}}}.$$

Therefore the continued fraction expansion of $\frac{92}{73}$, and indeed of any rational number, has one division for each line of the Euclidean Algorithm as applied to the number pair. Notice that, unlike the number $\sqrt{2}$, which is *not* a rational number, the continued fraction expansion stops.

It is standard when studying expansions of this kind to restrict attention to continued fractions in which all numerators equal 1. None the less, expansions that allow for other numerators have been investigated. Here, however, we shall confine ourselves to the normal type.

There is nothing to prevent us carrying out the same style of calculation for any positive irrational number a: we simply apply the Euclidean Algorithm to the pair of numbers $(a, 1)$. The difference is that in this case we shall never arrive at a remainder of 0, for if we did we could carry out the above process to express a as a terminating continued fraction which could be simplified to an ordinary fraction, showing that a was rational after all. A recurring pattern can emerge, however, in the continued fraction expansion, as we have already seen demonstrated with $\sqrt{2}$. It turns out that the irrationals that yield such a recurring pattern are exactly the irrational roots of quadratic equations with integer coefficients. As another example, let us look at $\sqrt{3}$.

The first step is to write the given number a in the form $a = n + r$ where n is a whole number and the remainder, r, is less than 1: this is why we began our first example by writing $\sqrt{2} = 1 + (\sqrt{2} - 1)$, as $1 < \sqrt{2} < 2$, so that here $n = 1$ and $r = \sqrt{2} - 1$. Taking $a = \sqrt{3}$ and following the pattern of calculation given by (1), we get:

$$\sqrt{3} = 1 + (\sqrt{3} - 1) = 1 + \frac{1}{\frac{1}{(\sqrt{3}-1)}}.$$

Following (2), we next need to express $\frac{1}{(\sqrt{3}-1)}$ in the form of $n + r$, that is to say, integer + remainder. This is achieved through first rationalizing the denominator:

$$\frac{1}{\sqrt{3} - 1} = \frac{1}{\sqrt{3} - 1} \cdot \frac{1 + \sqrt{3}}{1 + \sqrt{3}} = \frac{1 + \sqrt{3}}{3 - 1} = \frac{1 + \sqrt{3}}{2}.$$

Now, since $1 < \frac{(1+\sqrt{3})}{2} < 2$, we next write this as:

$$\frac{1 + \sqrt{3}}{2} = 1 + \left(\frac{1 + \sqrt{3}}{2} - 1 \right) = 1 + \frac{1 + \sqrt{3} - 2}{2}$$

$$= 1 + \frac{\sqrt{3} - 1}{2}.$$

We can now advance to the stage corresponding to (3) above:

$$\sqrt{3} = 1 + \cfrac{1}{1 + \frac{\sqrt{3}-1}{2}} = 1 + \cfrac{1}{1 + \frac{1}{\frac{2}{(\sqrt{3}-1)}}}.$$

Continuing in this way, we obtain:

$$\frac{2}{\sqrt{3}-1} = \frac{2}{\sqrt{3}-1} \cdot \frac{\sqrt{3}+1}{\sqrt{3}+1} = \frac{2\sqrt{3}+2}{2} = \sqrt{3}+1.$$

Writing this in the form $n + r$ yields:

$$\sqrt{3} + 1 = 2 + (\sqrt{3} - 1),$$

and so we arrive at:

$$\sqrt{3} = 1 + \cfrac{1}{1 + \frac{1}{2+(\sqrt{3}-1)}} = 1 + \cfrac{1}{1 + \cfrac{1}{2 + \frac{1}{(\sqrt{3}-1)}}}.$$

Since $\sqrt{3}$ is irrational, the process is unending but we have now the same remainder, $\sqrt{3} - 1$, as earlier in the process. It follows that the calculation has fallen into a recurring pattern in that the value of n will henceforth alternate between 1 and 2:

$$\sqrt{3} = 1 + \cfrac{1}{1 + \frac{1}{2 + \frac{1}{\ldots}}},$$

or in the more compact notation introduced earlier:

$$\sqrt{3} = [1, \dot{1}, \dot{2}]. \tag{4}$$

Similarly, we can calculate the following continued fraction expansions:

$$\sqrt{5} = [2, \dot{4}], \quad \sqrt{7} = [2, \dot{1}, 1, 1, \dot{4}].$$

The continued fraction expansion of a number is rich with information.

The fractions that result from terminating the expansion at any stage are called the *convergents* of the number a. They do indeed approach the number a as their name suggests, giving values that are alternately over- and underestimates of the exact value. They also enjoy other good properties that allow, for example, testing for irrationality of certain numbers. As mentioned above, recurring expansions with unit numerators arise only for numbers of very special type, although some other irrationals have continued fraction representations which have special patterns. For example, one application of the Wallis product mentioned in Chapter 7 is to show that:

$$\frac{4}{\pi} = 1 + \cfrac{1^2}{2 + \cfrac{3^2}{2 + \cfrac{5^2}{2 + \cfrac{7^2}{2 + \cdots}}}}.$$

Since we saw in Chapter 2 that you can turn any recurring decimal back into a fraction, this prompts us to ask whether or not we can travel in the reverse direction in this new context. For example, surely the number

$$\alpha = 1 + \cfrac{1}{1 + \cfrac{1}{1 + \cfrac{1}{1 + \cdots}}},$$

that is to say $\alpha = [\dot{1}]$, must be quite special. This number is called the *Golden Ratio* and can be extracted from its continued fraction expansion quite easily. The thing to spot is that what appears under the first division line is merely another copy of $\alpha = [\dot{1}]$ so that α satisfies the equation:

$$\alpha = 1 + \frac{1}{\alpha}. \tag{5}$$

Multiplying both sides of this equation by α reveals it to be a quadratic:

$$\alpha^2 = \alpha + 1 \Rightarrow \alpha^2 - \alpha - 1 = 0. \tag{6}$$

Solving this equation with the quadratic formula yields two solutions, one positive and one negative; clearly, it is the positive one that we require:

$$\alpha = \frac{1 + \sqrt{5}}{2} \approx 1.618.$$

This is somewhat anti-climactic as we have an unremarkable-looking answer. More will be revealed if we focus on the *properties* of the number rather than this expression for it. Another relationship for α arises through subtracting 1 from both sides of equation (5):

$$\alpha - 1 = \frac{1}{\alpha}. \tag{7}$$

Then taking reciprocals of both sides of (7) yields:

$$\alpha = \frac{1}{\alpha - 1}. \tag{8}$$

Bear this in mind as we consider a rectangle with sides of length α and 1 (Figure 1). If we slice the largest square possible from the rectangle of part (*a*), a 1×1 square as shown, the remaining rectangle has a long side of 1 unit and a short side of $\alpha - 1$ units. These two rectangles are in fact *similar*: if we look at the ratio of their sides we see that they are, respectively, $\frac{\alpha}{1}$ and $\frac{1}{(\alpha-1)}$ and equation (8) says that these are the same. Of course, since the smaller rectangle is the same shape as the original, applying the same process to the smaller rectangle (Figure 1(*b*)) leads to the same result, which can then be repeated ad infinitum.

The rectangle with this shape is called the *Golden Rectangle*. Its

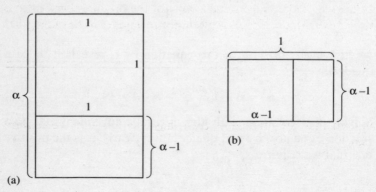

FIG. 1

special properties were a source of fascination for the Greeks and in the early sixteenth century a book, *De Divina Proportione*, was written on the subject by Pacioli. It is often said that the Golden Rectangle is the one whose proportions are the most pleasing to the eye and so it is favoured in design. I'm not sure about this: for example, the shape of the standard credit card is not golden, although it is quite close. None the less, I would expect that readers could find examples of the Golden Rectangle at work in wallpaper patterns, architecture, and such like.

This process of extraction of the largest integer-sided rectangle from a rectangle of sides of length a and 1 does correspond to the construction of the continued fraction representation of a. The first extraction corresponds to writing $a = n_1 + r_1$, where r_1 is the length of the shorter side of the remaining rectangle. The step where top and bottom of the remainder fraction $\frac{r_1}{1} = \frac{r_1}{(n_2 r_1 + r_2)}$ are divided by r_1 can be regarded as scaling the remainder rectangle so that the shorter side of length r_1 is now treated as a unit length. It was known to both the ancient Greeks and Indians that if we take $a = \sqrt{d}$ for an integer d then eventually two of the remainder rectangles are similar and so the continued fraction will be of recurring type, although this was proved rigorously only by Lagrange in the eighteenth century.

An even simpler geometric situation that gives rise to the Golden Ratio is to take a line segment and ask: what is the value of a such that, if a section of length a is removed, the ratio of a to the remainder is the same as that of the original segment to a itself? (See Figure 2.) Let us deem the remainder to be of length 1

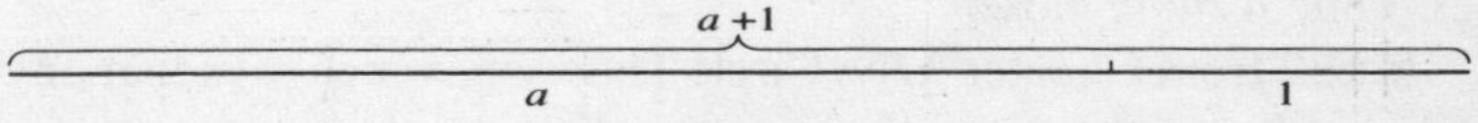

FIG. 2

unit so that the original segment has length $a + 1$. We require that:

$$\frac{a+1}{a} = \frac{a}{1} \Rightarrow 1 + \frac{1}{a} = a,$$

whereby we see that a does indeed satisfy equation (5) which characterizes α. In this context α is often referred to as the *Golden Section*.

The power of the pentagon

A third geometric instance where the Golden Ratio, α, arises quite strikingly is through the diagonal of a regular pentagon the side of which has unit length. Indeed, the length d of any diagonal of such a pentagon *is* the Golden Ratio. The pentagon with its diagonals is pictured in Figure 3. This is often imagined

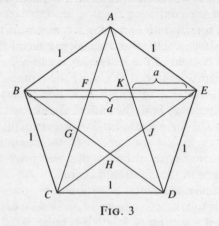

Fɪɢ. 3

as a powerful if not sinister symbol. The diagonals give the figure its strength and the hidden symmetries its mystery. Let us examine it more closely.

Recall from Chapter 3 the Circle Theorem, which says that any two angles at the circumference standing on the same arc of a circle are equal. A consequence of this is that, for any regular polygon with n sides, any angle of the type $\angle ACB$, where AB is a side and C another vertex of the polygon, equals $\left(\frac{180}{n}\right)^\circ$: see Figure 17 in that chapter. Applying this to our pentagon, we see that angles such as $\angle BAC$ and $\angle CAD$ all equal $\left(\frac{180}{5}\right)^\circ = 36^\circ$. In the same chapter we saw that the sum of the angles of a polygon is $(n-2) \times 180^\circ$ so that each of them has a measure of $\frac{n-2}{n} \times 180^\circ$.

In the case of a pentagon where n is 5, we see that an angle such as $\angle BAE$ is equal to $\frac{3}{5} \times 180° = 108°$.

Now the triangle ABK is isosceles as two of its angles, $\angle BAK$ and $\angle BKA$, both equal $72°$:

$$\angle BAK = \angle BAC + \angle CAD = 36° + 36° = 72°;$$

$\angle ABK = 36°$, and so

$$\angle BKA = 180° - \angle BAK - \angle ABK = 180° - 72° - 36° = 72°.$$

It follows that the lines AB and BK are both of unit length, and that the segment KE, which we denote by a, is related to the diagonal d by $d = a + 1$.

Next, the triangles ABE and AKE are similar as they have the same angles ($108°, 36°, 36°$) and so by taking ratios of corresponding sides we obtain $\frac{d}{1} = \frac{1}{a}$ or, what is the same, $ad = 1$. We thus have the equations:

$$d = a + 1, \quad ad = 1.$$

Multiplying the first by d and using the fact that $ad = 1$, we see that:

$$d^2 = ad + d = 1 + d$$
$$\Rightarrow d^2 - d - 1 = 0,$$

which is equation (6) for the Golden Ratio α; that is to say, $d = \alpha$. Therefore the diagonal of the pentagon has length equal to that of the Golden Ratio. Moreover, we have discovered other properties of the pentagon including the interesting one represented by the equation $ad = 1$. This is equivalent to the statement that the segment BK, which has length 1, is a Golden Section of the diagonal BE, as since $d = \frac{1}{a}$ we have:

$$\frac{d}{BK} = \frac{d}{1} = \frac{1}{a} = \frac{BK}{a},$$

and the statement that

$$\frac{d}{BK} = \frac{BK}{a}$$

says exactly that the section *BK* of the diagonal *BE = d* is a Golden Section.

In summary, the length of each diagonal of the pentagon is the Golden Ratio while the diagonals intersect one another in the Golden Section.

Fibonacci's Rabbits and the Golden Ratio

Fibonacci's Rabbit Problem dates back to the early thirteenth century. It introduced a sequence of numbers that are generated in so simple and natural a fashion that they were bound to arise again and again. Their persistent appearance in natural phenomena, especially situations involving growth, is none the less remarkable. Indeed, the original setting for the sequence is the following population problem.

We begin with a rule. Each pair of rabbits gives birth to another pair in the next generation and to a second pair in the generation after that, after which they are too old to reproduce any further.

The first generation consists of a single pair; the second generation therefore also consists of just one (new) pair, but in the third generation there will be two pairs born as each of the first- and second-generation pairs contribute. The fourth generation will have three pairs: two pairs will be the offspring of the pairs from the third generation while the other will be from the second-generation parents. The first dozen Fibonacci numbers, that is the numbers of pairs in each generation, then turn out as follows:

$$1, 1, 2, 3, 5, 8, 13, 21, 34, 55, 89, 144, \ldots$$

Can you see a pattern? You shouldn't be able to, not at least like ones we have seen in number sequences we have met up until now. There is no *simple* formula to relate f_n, the nth number in this sequence, to n itself (although there is a complicated one). This is nevertheless an easy sequence to generate because of the following observation. Consider f_n, the number of rabbits born into the nth generation. The only generations that contribute offspring to any particular generation are the two that precede it. Each pair born in generation number $n - 2$, of which there are f_{n-2} pairs, and each pair of the preceding generation numbered $n - 1$, which has a population of f_{n-1} pairs, contributes one pair to the nth generation. We can therefore say that:

$$f_1 = f_2 = 1, \quad f_n = f_{n-1} + f_{n-2}, \quad n \geq 3.$$

This certainly allows you to calculate the Fibonacci numbers very readily (check the first 12 values for yourself), although a method such as this is known as a *recursion*, not a formula—in order to find f_{100} you will need to find all the preceding members of the sequence first.

Where lies the connection with the Golden Ratio? It looks to be nowhere in sight. The Fibonacci sequence is clearly *not* a geometric sequence as the ratio of successive terms is not constant; this is easily checked. A little perseverance is rewarded, however. If you calculate the quotients $\frac{f_n}{f_{n-1}}$ for many values of n you will notice something quite remarkable. Although no two ratios are ever exactly equal, after a time they are all almost equal. If you are even more astute you will notice that the value they are all approximating is around 1.618..., the Golden Ratio! The Fibonnaci sequence behaves, in the long run, like a geometric sequence with common ratio α. Why should this be?

In fact, once you suspect that this is true, it turns out to be easy enough to explain and the recursion can be exploited in order to establish it. Taking $n \geq 3$, we begin with $f_n = f_{n-1} + f_{n-2}$. Divide through by f_{n-1} to obtain:

$$\frac{f_n}{f_{n-1}} = 1 + \frac{f_{n-2}}{f_{n-1}}.$$

We next write $f_{n-1} = f_{n-2} + f_{n-3}$ to get:

$$\frac{f_n}{f_{n-1}} = 1 + \frac{f_{n-2}}{f_{n-2} + f_{n-3}} = 1 + \frac{1}{1 + \frac{f_{n-3}}{f_{n-2}}},$$

where we have divided top and bottom by f_{n-2} to get the last equality. We continue, replacing f_{n-2} by $f_{n-3} + f_{n-4}$ and dividing top and bottom of the resulting fraction by f_{n-3}, to give:

$$\frac{f_n}{f_{n-1}} = 1 + \frac{1}{1 + \frac{1}{1 + \frac{f_{n-4}}{f_{n-3}}}}.$$

By continuing in this fashion we shall eventually reach a finite continued fraction consisting entirely of 1s as the final Fibonacci ratio is $\frac{f_2}{f_1} = \frac{1}{1} = 1$. We conclude that the ratios of successive Fibonacci numbers do correspond to the truncated continued fraction expansion of the Golden Ratio α as given in (5). The limiting value of this ratio is α itself and so the value of the ratio $\frac{f_n}{f_{n-1}}$ approaches α for large values of n.

For the sake of readers' curiosity I will finish this part of the discussion with the statement of a preposterous-looking formula for the nth Fibonacci number, for there is one:

$$f_n = \frac{1}{\sqrt{5}} \left(\left(\frac{1 + \sqrt{5}}{2} \right)^n - \left(\frac{1 - \sqrt{5}}{2} \right)^n \right). \tag{9}$$

This is really a fright the first time you see it: after all, there is no obvious reason why the monster on the right should turn out to be an integer at all, never mind the nth Fibonacci number. You will, however, spot the reassuring presence of the Golden Ratio in the formula. In fact, we see both the roots of the Golden Quadratic $x^2 = x + 1$; let us call these roots α, the Golden Ratio, and $\beta = \frac{1-\sqrt{5}}{2}$.

This formula may be verified by an inductive argument. We check that the formula works for $n = 1$ and 2 and then use the Fibonacci recursion to show that the formula holds at every stage. The argument is straightforward enough and makes use of the fact that both the numbers α and β have the special property that they satisfy the quadratic equation $x^2 = x + 1$. It does nothing, however, to explain how such a formula could have been discovered in the first place! Rest assured that there is a standard technique for finding formulae for all recursions of the Fibonnaci type which takes this problem in its stride.

What use is such an awkward formula in any case? It is of no use for calculating Fibonacci numbers—if you want to find f_{100} you are much better off using the recursion repeatedly than wrestling directly with the formula. It is of theoretical use, however. For instance, although I will not give the details here, it is a simple matter using the formula to prove that the ratio $\frac{f_n}{f_{n-1}}$ does indeed approach the Golden Ratio as n becomes larger.

The Papal Sequence

Next I have quite a different type of Fibonacci phenomenon for you. Begin with two letters J and P and let them be the first two 'words' in a sequence of words formed from the Fibonacci-style rule: each word in the sequence is formed by the *concatenation* of the two previous words into a single word. The sequence begins as follows:

$$J, \; P, \; JP, \; PJP, \; JP^2JP, \; PJPJP^2JP, \; JP^2JP^2JPJP^2JP, \; \ldots,$$

where P^2 stand for PP. I first chanced upon this sequence in a frivolous way inspired by the naming of Pope John Paul I in 1978: that pontiff took his name from those of his two predecessors in this fashion. The above *papal sequence* is the one that would result if all his successors had felt obliged to follow his lead in this matter. However, I have since been assured that this sequence arises naturally in many diverse fields, such as the theory of abstract languages in computer science and the study of crystals. I will give a description of just some of the interesting facets of this sequence of words with an eye to a reasonable description of the name P_n of the nth pope.

If we begin numbering the sequence by taking the first word to be JP (regarding the J and P as generators), we can easily see that the number of Js in the nth word P_n is the nth Fibonacci number, f_n, while the number of Ps in P_n is actually f_{n+1}; the length of P_n is therefore $f_n + f_{n+1} = f_{n+2}$. It also not hard to see that if $m \leq n$ then P_n ends in P_m. Since the words of the papal sequence always end in JP, therefore, and never begin in P^2 (they start alternately with JP and with PJ), we see that no P_n can contain two successive Js or three successive Ps.

Denote the reverse of P_n by P_n^*. We can define an infinite sequence of letters A by considering the reverse of the papal sequence:

$$A = PJP^2JPJP^2JP^2JP\ldots$$

This makes sense, as the reverse of the papal sequence is *stable* in that for any integer k the final k letters of the words in the papal sequence are always the same from some point in the sequence onwards.

If we could generate the sequence A, we could find the name P_n: we would just take the first f_{n+2} letters of A and that would give us P_n^*.

It is easier to continue by coding the letters of A as 0, 1, and 2, with 0 replacing J, 1 replacing P, and 2 standing for P^2. The sequence A therefore consists of 1s and 2s separated by 0s; moreover, we know that 00 never occurs and that A begins with 1. Given these observations, we could reconstruct A if we just knew how to generate the pattern of 1s and 2s. Let B be the sequence derived from A by deletion of all the zeroes. It turns out that B can be generated using two simple rules.

Let B_0 be the sequence consisting only of the symbol 1. We construct further sequences B_1, B_2, ... using the following *rewriting rules*:

$$1 \to 12, \quad 2 \to 122.$$

These mean: whenever we see a 1 we replace it by 12; wherever we see 2 we replace it by 122. The first four of the B_is are then:

$$B_1 = 12, \quad B_2 = 12\ 122, \quad B_3 = 12\ 122\ 12\ 122\ 122,$$
$$B_4 = 12\ 122\ 12\ 122\ 122\ 12\ 122\ 12\ 122\ 122\ 12\ 122\ 122.$$

The sequences B_i carry the information that allows you to recover the name of Pope $2i + 1$. For instance, for P_7 we need $B_3 = 121221122122$. The recipe is as follows:

Reverse B_3, insert 0 at the beginning and between every pair of symbols 1 and 2, and finally, revert to J, P, and P^2 to recover the name:

$$B_3^* = 2212212122121 \to 0202010202010201020201010201$$

and so

$$P_7 = JP^2JP^2JPJP^2JP^2JPJP^2JPJP^2JP^2JPJP^2JP.$$

The five regular solids and the Golden Ratio

We close with an example from Antiquity, although the link with the Golden Ratio was a product of Renaissance mathematics.

The three-dimensional analogue of a regular polygon is a *regular polyhedron*, which is a figure bounded by identical regular polygons for which the same number of faces meet at any corner. For example, a cube is a regular polyhedron with square faces. It is of course possible to have regular polygons with any number of sides but regular polyhedra are much rarer. Indeed, there are only five of them.

How many polyhedra could there conceivably be with regular (that is to say equilateral) triangles for faces? At each corner of such a polyhedron there could be three, four, or five triangles meeting but never six, as this would give a combined angle of $6 \times 60° = 360°$ and the corner would be flat. (Obviously, more than six equilateral triangles at a vertex is out of the question.)

Using squares, we have already seen that we can have three squares at each corner giving a cube, but once again four squares would result in a flat vertex and more than four is impossible.

The internal angle of a regular pentagon is 108°, so it seems that it might be possible to construct a regular polyhedron with identical pentagons for faces, with three meeting at each corner, but no more than three. Hexagonal faces are impossible, as the internal angle is 120° and so three meeting at one corner could happen only in a plane, while more is impossible. For polygons with seven or more sides there is clearly no hope, as the internal angles of these polygons exceed 120°.

Are these five possibilities realizable? Let us examine the three cases where equilateral triangles are involved. A polyhedron where three equilateral triangles meet at each corner is certainly familiar enough: it is the *tetrahedron* or equilateral triangular pyramid (Figure 4). Four meeting at a corner is also possible, and in fact such a solid can be generated from a cube: join the centres

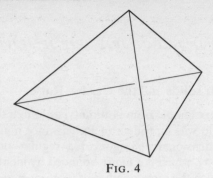

FIG. 4

of the faces of a cube that share a common edge and the result is an *octahedron* (see Figure 5). We say the octahedron is the *dual polyhedron* to the cube. (Duality is a two-way street: if we join the centres of the faces of an octahedron where those faces have a joint edge, the result is a cube.)

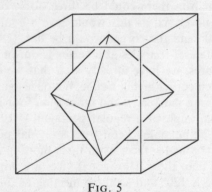

FIG. 5

Those of you alert by now to the way mathematicians work might try to do something more with this duality idea. Perhaps we get a regular polyhedron by taking the dual of the tetrahedron? The answer is yes but, disappointingly perhaps, the tetrahedron is *self-dual*: joining the centres of the faces of a tetrahedron gives us only another tetrahedron inside it (Figure 6). Keep that idea in mind, however. Although the five regular solids

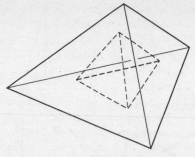

FIG. 6

were mathematical objects that took pride of place in the works of Euclid, it was Leonardo da Vinci's friend Luca Pacioli who saw a clever way to construct a regular solid where five triangles meet at each vertex. One simply has to consider three intersecting Golden Rectangles as in the picture that appears in John Stilwell's outstanding book, *Mathematics and its History* (see Figure 7). The 12 corners form a solid that has 20 triangular faces with five faces at each corner. I have drawn the five equilateral triangles associated with one particular corner explicitly in the picture. Since there are 12 corners, each associated with five triangles, and since each triangle involves three of the corners as its vertices, we see that the enclosing solid will have $\frac{(12 \times 5)}{3} = 20$ triangular faces.

It remains to check that all these triangles truly are equilateral: in fact, the side length of a typical triangle ABC is 1, as two applications of Pythagoras reveals. Bear in mind that for each rectangle its shorter sides are of length 1 unit while the longer side is of length α, the Golden Ratio, and that α has the property that $\alpha^2 = 1 + \alpha$. Let M be the midpoint of the side BC and let D be the point where the two rectangles whose corners define the triangle ABC meet, as shown in Figure 7. First, by Pythagoras,

$$AM^2 = MD^2 + AD^2.$$

Now $MD = \frac{(\alpha - 1)}{2}$ and $AD = \frac{\alpha}{2}$. Hence:

$$(\alpha - 1)^2 = \alpha^2 - 2\alpha + 1 = \alpha + 1 - 2\alpha + 1 = 2 - \alpha.$$

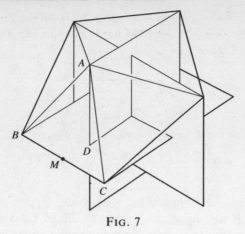

Fig. 7

We thus obtain:

$$AM^2 = MD^2 + AD^2 = \left(\frac{\alpha - 1}{2}\right)^2 + \left(\frac{\alpha}{2}\right)^2 = \frac{2 - \alpha}{4} + \frac{1 + \alpha}{4} = \frac{3}{4}.$$

Using Pythagoras a second time, we thus arrive at:

$$AB^2 = AM^2 + BM^2 = \frac{3}{4} + \frac{1}{4} = 1,$$

and so AB has length one unit, as do all the lengths of the sides of the triangles in the picture, thus showing that the triangles are indeed equilateral. The regular solid formed from 20 equilateral triangles is called the *icosahedron*.

To obtain the fifth and final regular solid, we return to the idea of duality. The cube has six faces so that its dual, the octahedron, has six corners, one for each face of the cube, and the faces are equilateral triangles as the cube has three faces meeting at each of its corners. In the same fashion, the dual of the icosahedron, known as the *dodecahedron*, has one corner for each face of the icosahedron giving it 20 corners in all. Since five faces meet at each corner of the icosahedron, this causes the faces of the dual to have five sides each so that the dodecahedron is the regular solid with pentagonal faces. Each face of the icosahedron is joined to three others so that three faces of the dodecahedron meet at each of its corners. The final dual pair of regular solids is pictured in Figure 8.

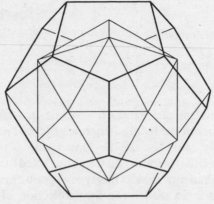

FIG. 8

We see therefore that there are five regular solids, although it is still conceivable that there could be more—for example, how do we know there is not another regular solid with five triangular faces meeting at each vertex with a different number of edges and faces than the icosahedron? A reason why no such object exists will be given in the final chapter.

10

Networks

The ninth problem of Chapter 6 was one in which we showed that the network of the Königsberg bridges could not be traversed once and only once because the underlying network had too many *odd nodes*, that is nodes that were connected to an odd number of edges. We shall now look at this type of problem in general.

By a *network* we shall mean any collection of *nodes* (sometimes called *vertices* and at other times simply *points*) and *edges* running between the nodes. We allow for several edges running between the same pair of nodes (*multi-edges*), and *loops*, edges that begin and end at the same node, are not forbidden. Moreover, in general a network may not necessarily consist of a single connected piece but may have a number of *components*. The edges of the network may cross each other, and indeed if there are many edges that can sometimes be unavoidable. If, however, the network can be drawn without edges crossing, then it is called a *planar* network. All these features can be seen in Figure 1.

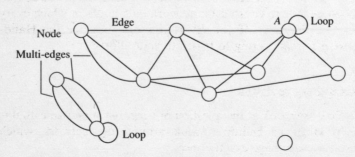

Fig. 1

There is one observation that holds for the most arbitrary network. By the *degree* of a node we mean the number of edges that are incident with it (loops counting double). For instance, the node A in the above example has degree 6. If we sum all the degrees of all the nodes of a network, we obtain a number s that is always exactly twice e, the number of edges, because each edge contributes 2 to the overall count, once at each of its two ends. In the example in Figure 1 we can count 18 edges all told, and summing their degrees by components we get:

$$(3 + 5 + 6 + 5 + 4 + 3 + 2) + (3 + 5) + 0 = 28 + 8 + 0$$
$$= 36 = 2 \times 18,$$

in accord with our general observation that $s = 2e$.

Let us write s_e for the sum of the degrees of all the even nodes (that is the nodes of even degree), and s_o for the corresponding sum for the odd nodes. We have:

$$s_e + s_o = s = 2e.$$

This gives $s_o = 2e - s_e$. Now s_e, being a sum of even numbers, is itself even, as is $2e$, so it follows that s_o is even as well. Since s_o is a sum of odd numbers, this is possible only if the number of numbers in the sum s_o is even; that is to say, the *actual number of odd nodes in the network must itself be an even number*. We deduce that any network has an even number of nodes of odd degree. (The above example has 6 odd nodes.) This fact, sometimes called the *Hand-Shaking Lemma*, is quite useful: it is certainly important to appreciate that it is true. For instance, it tells us that it is impossible to construct a network on 5 nodes in which every node has degree 3. If you try it you will soon see the Hand-Shaking Lemma acting to frustrate your efforts.

Königsberg revisited

We shall now look at the question of *traversing* a network, that is the problem of finding a walk around the network which traverses every edge exactly once.

The argument we presented concerning the Königsberg bridges

shows that, for a network N to be traversable, N can have at most two odd nodes which must then lie at either end of the traversing walk. If we demand more and ask for a traversing *circuit*, that is a traversing walk which begins and ends at the same node, the pairing argument given in Chapter 6 shows that this will be impossible unless all the nodes are even. (A circuit must arrive and leave any node an equal number of times and so there must be an even number of edges connected to the node.) It turns out that this necessary condition is also sufficient to traverse a connected network N: N is traversable if (and only if) all of its nodes are even. (Obviously, we cannot hope to traverse a network that has several components.)

Can we find a method of actually doing this? Will anything work? Perhaps if we have such a network N and we walk around it in any fashion, using a new edge at every juncture, we shall eventually end up back where we began having used up all the edges? This totally simple-minded approach will not always work—if you are not careful you can become stuck.

Take the next example (Figure 2). This is a connected network

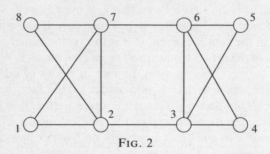

FIG. 2

in which every node is even. However, if we start at 7 and our walk begins $7 \rightarrow 6 \rightarrow 3 \rightarrow 2$, we have landed ourselves in trouble. If we imagine burning our bridges after we have walked over them, we see that upon arrival at 2 the remaining network has split into two components and that we have managed to strand ourselves on the left-hand side. However, this is the only difficulty that can arise and it is easily avoided. We do not have to be very clever when we construct our walk—we don't have to think two steps ahead—we only need to avoid taking a step that splits the remaining network in two. We can indeed give an *algorithm* that will do the job, that is to say a mechanical procedure that avoids the necessity of true judgement or intelligence.

Begin at any node traversing the network any way you wish, but:

1. Draw a picture of the network and erase any edge that you use and any node that has had all of its edges traversed.
2. At each step use an *isthmus*, that is an edge connecting two otherwise disconnected parts of the remaining network, only if there is no choice.

You should have no difficulty traversing the above network now, beginning at any node that you please. (Note that the third step, $3 \rightarrow 2$, of our failed walk violates Rule 2 by crossing an isthmus.)

If the connected network has odd nodes it must, by the Hand-Shaking Lemma, have at least two of them. If it has more than two we know that there is no traversing walk; but what if there are exactly two odd nodes? Can the above method still cope to produce a traversing walk even if it is not a circuit? The answer is yes, as I shall now explain.

We can traverse the network, beginning at either of the two odd nodes and ending at the other. Call the two odd nodes A and B respectively. Draw another edge e on the network from A to B. In this augmented network all of the nodes are now even so our previous algorithm allows us to find a traversing circuit, starting from B, and we can also insist that the new edge e that we adjoined is the first one that we use. However, this circuit will then consist of walking from B to A along e and the remainder must be a traversing walk of the original network which begins at A and ends at B. Try your hand on the next example, which has two odd nodes, as indicated in Figure 3.

A proof that the above algorithm always works (I have simply said that it does) can be found in any serious book about networks and *graph theory*, as the study of networks is often called. The proof is not long or difficult, but is slightly awkward if you insist on justifying every detail, which many books prefer not to do so as not to mar the simplicity of the underlying idea.

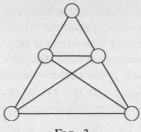

FIG. 3

Crossed wires: can we avoid them?

The second type of puzzle problem involving networks is where you are asked to draw a network involving certain connections in such a way that the edges do not cross. The standard example is where there are three houses that are each to be connected to gas, water, and electricity outlets. To minimize the possibility of one service cutting the supply of another during maintenance, it would be best if the connections could be made without any one supply line ever crossing over the top of another. But can it be done? A near-successful failure is shown in Figure 4.

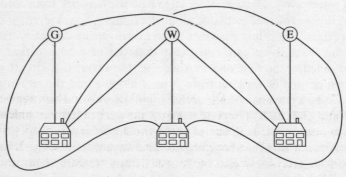

FIG. 4

You should be able to do equally well, but no better. How can you *prove* this to be impossible? How can we be sure that there is not a clever way of doing it that we just have been unable to spot?

The difficulty lies not so much in that the problem is genuinely complicated but that, whatever the justification, at some stage one needs to assert that it is obvious that one edge now has to cross another because it must pass from the inside to the outside of some figure which is inevitably created by the other edges. There is nothing wrong with this except that it is very difficult to justify rigorously as even simple closed curves are difficult to deal with in full generality.

In fact, there are two fundamental networks that are not *planar*; that is to say, they cannot be drawn without at least one pair of edges meeting at a point that is not a node of the network. The first we have already mentioned, $K_{3,3}$, the network that arises from connecting every member of one set of 3 nodes to another set of 3 nodes. The second is K_5—the so-called complete network on 5 nodes: the *complete network* on n nodes has exactly one edge running between each pair of nodes (Figure 5).

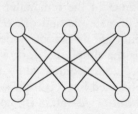
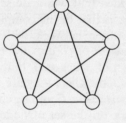

Fig. 5

The importance of $K_{3,3}$ and K_5 lies not only in their not being planar but in the fact that every network is planar unless it 'contains' a copy of one of these two forbidden networks (in a sense that can be made precise). This theorem, which is difficult to state precisely and to prove, was demonstrated by Kuratowski in 1930. Before discussing planarity any further we shall pause to clarify aspects of the general situation.

We need only concern ourselves with networks that do not have loops or multiple edges: we shall call such networks *simple networks*. The reason for this is that if a network N is planar then the underlying simple network, which is obtained by deleting all loops and coalescing any multiple edges between two nodes into a single edge running between them, is also planar. Conversely, if the underlying simple network of a network is planar, then we can replace any single edge of the underlying simple network by the required number of multiple edges and adjoin any number of loops we wish to the picture without violating the planarity; therefore if the underlying simple network is planar then so is the network.

We have already solved a problem about simple networks in Chapter 6, where we saw that at any party there were at least two people with the same number of friends at the party. We can recast this question as one about simple networks: draw a network that has one node for each person and for which two nodes are connected by an edge if the people they represent are friends. What the argument of Chapter 6 then demonstrated was that in any simple network there must be at least two nodes of the same degree.

An idea that will arise several times in the remainder of this chapter is that of the complement of a simple network G. Let G be a simple network with N denoting its set of nodes. The *complement* of G, denoted by $\overline{G}$, is the simple network with the same set of nodes as G, but with two nodes connected by an edge in $\overline{G}$ exactly where they are not connected in G. It follows that if we superimpose G and its complement $\overline{G}$ we obtain the complete network on the set of nodes N. Taking the complement of the complement of G of course gives us G back again: $\overline{\overline{G}} = G$: see Figure 6.

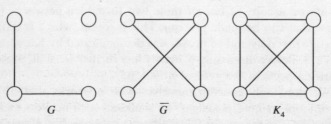

G $\overline{G}$ K_4

FIG. 6

In the eighth problem of Chapter 6 we saw that at a party of six or more there is always a triangle of mutual acquaintances or a triangle of mutual strangers. This also can be formulated neatly in terms of networks and their complements as the networks of acquaintanceship and of non-acquaintanceship are mutual complements of one another: mutual acquaintanceship is represented by edges in G while the edges of $\overline{G}$ denote non-acquaintanceship. What the problem is really asking us to show is that, for any simple network G with at least six nodes, either G contains a copy of K_3 (that is, a triangle of edges) or its complement $\overline{G}$ does. This can be seen in the previous example where $\overline{G}$ has a requisite triangle although G does not. (It is of course perfectly possible for both G and $\overline{G}$ each to contain numerous triangles.)

An instructive example arises if we look at the network G on five nodes which can be drawn as a regular pentagon. This we have already met in Chapter 6 because it provided, when thought of as representing five dinner guests around a table, an example of a party where there was no triangle of acquaintanceship or non-acquaintanceship. The pentagon G contains no triangle. If we draw $\overline{G}$ in the obvious way though, the picture that results is not so helpful (see Figure 7). The network $\overline{G}$ looks more

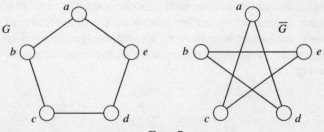

FIG. 7

complicated than G: it does not even look planar as it has edges crossing in many unwanted places. On close inspection, however, we can see that $\overline{G}$ has really exactly the same network structure as G, and in particular it also lacks triangles. To clarify this, we need only order the way in which we list the nodes around the outside of the pentagon: instead of having the anti-clockwise order *a*, *b*, *c*, *d*, *e*, we order them *a*, *c*, *e*, *b*, *d*; the picture of $\overline{G}$ is now an ordinary pentagon (and *G* would now take on the star-shape)—see Figure 8.

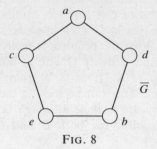

FIG. 8

It has turned out, therefore, that the networks *G* and $\overline{G}$, although representing different relationships, are the same when only their network structure is taken into account. Mathematicians have a word for this: we say that two networks are *isomorphic* if they can be represented by the same picture. This amounts to saying that there is a one-to-one correspondence of the nodes of the two networks such that two nodes are *adjacent* in the first graph (meaning that they are connected by an edge) if and only if the corresponding nodes in the second graph are also adjacent. In general, it can be difficult to tell whether or not two networks are isomorphic. For example, the two pictures in Figure 9 both represent $K_{3,3}$. A suitable correspondence of the nodes that shows this is:

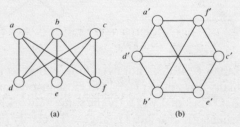

FIG. 9

$$a \to a', b \to b', \ldots$$

I leave it to the reader to check that two nodes are adjacent in the first network if and only if their dashed counterparts in the second network are also adjacent.

You can imagine the difficulties involved when dealing with very complicated networks. However, it is sometimes easy to spot that two networks are not isomorphic: one just has to see some essential difference. For instance, if the networks do not have the same number of nodes or the same number of edges, then they are certainly not isomorphic. The difference could be more subtle. One could have a node of degree 4 connected to a node of degree 6, while the other does not: if this were the case they would not be isomorphic. For example, spot an essential difference between the two networks shown in Figure 10. Both networks have two nodes of degree 3, but in the second they are mutually adjacent, which is not the case for the first, and because of this it is not possible to label the nodes of the networks

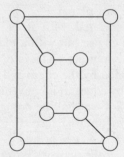
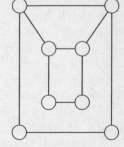

FIG. 10

$a, b, c, \ldots$ and $a', b', c', \ldots$ in a way that respects adjacency. If, however, you have an example where no such difference in structure is apparent, the problem of whether or not the networks are isomorphic can be very tedious.

We can now clarify what we mean by planarity. A network is planar if it can be represented by a *plane network*, that is a picture in which edges do not cross. The non-acquaintanceship network above is planar because it can be represented as a pentagon. In other words, it does not follow that a network is not planar simply because the first picture that you happen to draw of it has several edges crossing. For example, the network K_4 may be naturally drawn as on the left in Figure 11, but is easily seen to

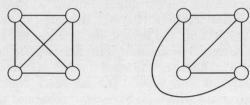

FIG. 11

be planar none the less, as the alternative picture on the right shows.

There is a special formula that applies to planar networks and this can be used to clarify to some extent the idea that a network is not planar if it has too many edges: in particular, this leads to explanations as to why $K_{3,3}$ and K_5 are not planar.

Any network has two numbers associated with it: n, the number of nodes, and e, the number of edges; but with a planar network we can associate a third, f, the number of faces of the plane figure, a *face* being a region bounded by the edges and not containing any smaller region bounded by edges. It is convenient, although it makes little difference to what we shall be doing, to count the outside of a plane network as another face. For the plane copy of K_4 above, we see that $n = 4$, $e = 6$, and $f = 4$. Here f_4 is its outside face. Clearly, however the network is drawn, the numbers n and e will remain the same, but this is not so clear as regards f. It *is* true, however, since for any connected plane network the three numbers are related by a very simple equation:

$$n + f = e + 2. \tag{1}$$

In the example of K_4 we see that this is satisfied: $4 + 4 = 6 + 2$. The reason why this relationship persists in any connected plane network can be seen by imagining the network to be drawn one edge at a time, adding any new nodes as necessary, and noting that the sums on both sides of the equation always go up and down together. When we draw the first edge we have a plane network with $n = 2$, $e = 1$, and $f = 1$ (just one unbounded face outside of the network), and so the equation holds. As we add more edges there are two cases to examine. Since the final picture is connected, it can be assembled, edge by edge, without the need of ever drawing another edge which is not connected to any of the nodes already drawn, although we may need to introduce one new node when drawing a new edge. There are two cases:

1. An edge is drawn which involves introducing a new node that does not lie on an existing edge, as in Figure 12. In this situation both n and e increase by 1 for each edge added, so equation (1) remains in balance: in the original network (1) took the form $4 + 3 = 5 + 2$, which changed to $6 + 3 = 7 + 2$

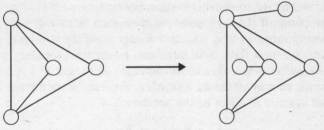

FIG. 12

for the network on the right, which was formed by adjoining two new edges of the type described.

2. An edge is drawn which involves no new nodes (Figure 13). This will cause both e and f to increase by 1, as the new edge will split an existing face (perhaps the outside face) in two. This will add 1 to each side of the equation.

In the example shown in Figure 13 two edges of this type are added, one of which divides an internal face in two while the other does likewise to the outside face. The formula (1) then passes from $6 + 3 = 7 + 2$ to $6 + 5 = 9 + 2$, but it still remains in balance.

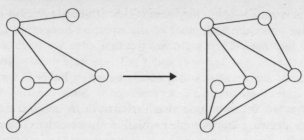

FIG. 13

This verifies the so-called *Polyhedral Formula* for plane networks. We can now exploit this connection between n, f, and e to identify some special properties that a planar network enjoys.

Suppose we have a simple plane network which is *connected*, that is to say consists of a single component, with f faces. Let us call the number of edges bounding a face the *face number* of that face and let us denote this sequence of numbers by $F_1, F_2, \ldots, F_f$. If we sum all the face numbers then each edge will have been counted at most twice, since every edge is on the boundary of no more than two faces. (An edge can be on the boundary of just one face, as shown by one of the edges in Figure 13.) It follows that the sum of all the face numbers is no more than twice e, the total number of edges of the network:

$$F_1 + F_2 + \ldots + F_f \leq 2e.$$

Now for any simple plane network with at least three edges, every face is bounded by at least three edges as we could have a two-edged face only if we allowed multiple edges, and a one-edged face would be bounded by a loop, both features that are forbidden in simple networks. In other words, each of the face numbers is at least 3 so that:

$$\underbrace{3 + 3 + \ldots + 3}_{f \text{ times}} \leq F_1 + F_2 + \ldots + F_f.$$

Combining the previous two facts yields:

$$3f \leq 2e \Rightarrow f \leq \frac{2}{3}e.$$

Combining this with our Polyhedral Formula (1) gives:

$$e + 2 = n + f \leq n + \frac{2}{3}e$$

$$\Rightarrow e + 2 \leq n + \frac{2}{3}e \Rightarrow \frac{1}{3}e + 2 \leq n.$$

In other words, by multiplying throughout by 3, we see that in any connected, plane, simple network with more than two nodes:

$$e \leq 3n - 6.$$

This disqualifies K_5 straight away from entry into the realm of planar networks as it has too many edges: for K_5, $e = 10$ and $n = 5$ so that e is greater than $3n - 6 = 9$.

However, $K_{3,3}$ momentarily escapes our net, as here we have $e = 9$ while $n = 6$ and so the $e \leq 3n - 6$ rule is respected by $K_{3,3}$. It does, however, surrender to the following argument similar to the one you have just read. Since the edges of $K_{3,3}$ always run between the two sets of three nodes (see Figure 9), it follows that any circuit in the network must have even length: in particular, there are no triangles, so in any plane representation of $K_{3,3}$, should one be possible, each face must be bounded by at least four edges. This gives us a stronger statement than before: namely $4f \leq 2e$; that is, $f \leq \frac{e}{2}$. Combining this with the Polyhedral Formula (1) shows that in any plane representation of $K_{3,3}$ we must have:

$$n + \frac{e}{2} \geq e + 2 \Rightarrow \frac{e}{2} \leq n - 2 \Rightarrow e \leq 2n - 4.$$

However, $K_{3,3}$ does not clear this hurdle, as $e = 9$, which is greater than $2 \times 6 - 4 = 8$.

The Polyhedral Formula is a very fundamental fact of mathematics. It has allowed us to show that neither K_5 nor $K_{3,3}$ can be drawn without at least one pair of edges crossing. The reason it is given the name that it has is that it applies to *polyhedra*, that is to solid figures bounded by plane polygonal surfaces, provided they are *convex*, i.e. the polygonal faces meet at angles less than a straight angle, such as the regular polyhedra of the previous chapter. For example, the dodecahedron has $n = 20$ corners, $e = 30$ edges, and $f = 12$ faces and so satisfies the formula $n + f = e + 2$. I invite you to check that the other four regular solids do as well.

If we accept the Polyhedral Formula, it is then easy enough to show that it is possible to answer a question of the kind raised at the close of the previous chapter: is it possible to have another regular solid that has five equilateral triangles meeting at each corner but with different numbers of edges and faces than the icosahedron? The answer is a firm 'no', because the Polyhedral Formula does not admit it. Suppose we have such a regular solid with n corners, e edges, and f faces. Counting all the edges at each corner gives $5n$ and since this will count each edge twice we see that $5n = 2e$. Similarly, counting edges by faces we see that, since each face has 3 edges and each edge lies on 2 faces, $3f = 2e$. Multiplying the Polyhedral Formula through by $3 \times 5 = 15$ gives:

$$15n + 15f = 15e + 30 \Rightarrow 6e + 10e = 15e + 30.$$

From this we see that e *must* be 30, and in consequence, n has to be $\frac{2}{5}e = 12$ and f is $\frac{2}{3}e = 20$: no other values are possible.

Bigger parties and bigger cliques

We now return to a type of problem first introduced in Chapter 6. How large a party does it take to ensure that there is a group of four mutual friends or a group of four mutual strangers? Cast in the language of networks, we ask: how many nodes does a simple network G require in order to ensure that either G or $\overline{G}$ contains a copy of K_4? It has been proved that this number is in fact 18. I cannot show that here. What I can show, however, is that the number does exist, and I present an argument to show that it is no more than 63. This may not sound too impressive, but remember that it is not obvious that the number must exist at all. The argument given is a very interesting one and is the basis of a proof of Ramsey's Theorem, which applies to collections much more general than the ones we are considering; it even has useful interpretations in cases involving infinite sets. In particular, the argument I give here can easily be extended to show that Ramsey numbers always exist—that is to say that, given a number n, there is a number N such that a simple network with N nodes or its complement has a copy of K_n inside of it. In Chapter 6, Question 8, we saw that if $n = 3$ then $N = 6$, the first interesting case. Now I shall show that for $n = 4$ the value of N is no more than 63.

It is easier to consider the network G and its complement to be superimposed, giving a copy of a complete network. Colour the edges white or black according as the edge lies in G or in its complement. What I shall show is that, provided the network has 63 nodes, it must contain a *monochromatic* copy of K_4; that is to say, there is some set of four nodes such that the edges passing between the nodes are all white, or they are all black.

Suppose then that our network (or our party if you like) has at least

$$1 + 2 + 2^2 + 2^3 + 2^4 + 2^5 \text{ nodes.}$$

If you remember your geometric series formula from the very first problem of Chapter 1, you will see that this number is $2^6 - 1 = 63$. (The number is not really important in what follows: it is chosen, as you will see, only to make sure that we have a sufficiently large supply of nodes to carry out the following procedure.) Focus on any one node—A_1 say—and proceed as follows (see Figure 14). Of all the edges leading from A_1 (there are at least 62 of them, of course, because we are using the complete network), at least half will be of one particular colour, call it C_1. (C_1 is either white or black.) Consider all the nodes connected to A_1 by an edge of colour C_1, and call this set of nodes S_1. There are at least

$$\frac{1}{2}(2 + 2^2 + 2^3 + 2^4 + 2^5) = 1 + 2 + 2^2 + 2^3 + 2^4 = 31$$

of these nodes. Let us choose one and call it A_2.

At least half of the edges from A_2 leading to the other nodes *in* S_1 are of the one colour; call this colour C_2 (which may or may not be the same as C_1). Let S_2 be the collection of these nodes. We note that S_2 is entirely contained in S_1 and itself has at least

$$\frac{1}{2}(2 + 2^2 + 2^3 + 2^4) = 1 + 2 + 2^2 + 2^3 = 15$$

members. Choose a member A_3 of S_2.

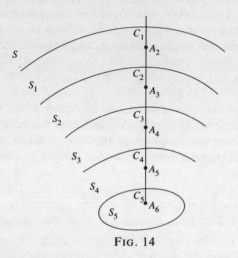

FIG. 14

We carry out this process five times, giving us nodes $A_1, A_2, \ldots, A_6$ and a collection of sets

$$S_1, S_2, S_3, S_4, S_5,$$

each of which is contained in the one before it, as indicated in Figure 14. The initial number of nodes was chosen in order to guarantee that we can carry out this process at least five times—the sets S_3, S_4, and S_5 will have at least 7, 3, and 1 member respectively. How does all this help? We need one subtle observation now to settle the question. The next paragraph has the key idea, but it requires some thought.

Consider the list of nodes A_1, A_2, A_3, A_4, A_5. Look at any member of this list, A_3 say. All the edges from A_3 to the members of the set S_3 are of the same colour. Now A_4, A_5, and A_6 *are all in* S_3, so all of the edges from A_3 to the members of the list that follow A_3 all have the same colour. This argument applies to A_1–A_5: each of the A_is has a colour associated with it, C_i, the colour of the edges leading from it to all the members of the list that follow it. Now there are only two colours available, black and white, and so at least three of $A_1, \ldots, A_5$ have the same colour (white say) associated with them. Choose such a group of three together with A_6: now every edge between these four nodes must be white, and so we have found our monochromatic copy of K_4— or, if you prefer, our clique of four mutual acquaintances.

Machines and languages

Our final topic involves looking at networks from a totally different perspective: as machines. The main ingredient of an *automaton* is a network in which the nodes are traditionally called *states*. Among the nodes there is an *initial state* and a number of *accepting states*. (There may be more than one of these, and the initial state may also be an accepting state.) At any one time an automaton $\mathcal{A}$ is in some state and may be acted on by an *input*, denoted by a letter of some set known as the *alphabet*, which has the effect of sending the automaton from one state to another. After a string of letters (a *word*) w acts on $\mathcal{A}$, the automaton will either be in an accepting state or not, as the letters of w take the automaton through some succession of states. We say the word w is *accepted* by the automaton, or is *recognized* by the automaton, if it leaves it in an accepting state. If not, w is rejected, and we say that the word is not part of the *language* recognized by the automaton.

If you like to indulge in anthropomorphism, you can think of the states of $\mathcal{A}$ as *moods* with the accepting states representing the machine's good moods and the other states its bad moods. It wakes up in its initial mood (which may be good or bad, depending on that individual machine) and the inputs it is subjected to leave it either in a good or a bad mood. If it finishes in a good mood then it accepts the word, but if the word puts it into a bad mood, then it rejects it.

For example, let us have a simple alphabet $A = \{a, b\}$. This will always suffice for our purposes, and indeed for most theoretical work two letters are enough. In Figures 15–17 the initial state is labelled i and the accepting states are shaded. The arrows on the edges indicate how a letter changes the automaton from one state to another.

The automaton depicted in Figure 15 recognizes a word provided that it contains at least one *b*. A word consisting of only *a*s never takes it out of the initial state. Once the machine

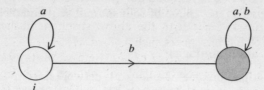

FIG. 15

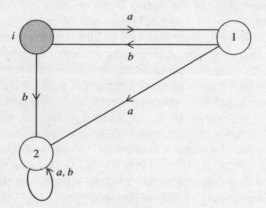

FIG. 16

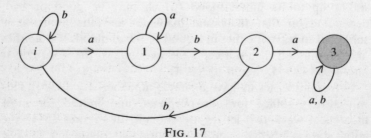

FIG. 17

sees a *b* it is happy, and it stays in its happy mood (the accepting state) no matter what it sees after that.

The next machine is not so easily pleased (Figure 16). This fellow will recognize a word only if it consists of a string of *ab*s, even the empty word (a string of zero *ab*s). For example, the word *abababab* will cause the machine to go from its initial state (which is also its only accepting state) to state 1 and back again to *i* four times. Since with the above string it finishes at the accepting state it recognizes such a word. However, as soon as it can tell it has not got a string of *ab*s it moves to its *sink* state 2, from which it will not budge. This will happen if you begin your word with *b*, or if your word ever has two successive letters the same. Either of these occurrences is enough to offend the machine as it will know that it is being offered a word that is not in its language, after which it totally loses interest.

As a third example, see if you can find the language accepted by the little automaton shown in Figure 17. This machine accepts a word when and only when that word contains the *factor aba*; for instance, *baabaa* is accepted while *abba* is not. In fact, this automaton is the smallest one that can be designed to accept this particular language.

The theory of automata is immense and has an algebraic theory all of its own which forms a part of the subject known as *algebraic semigroup theory*. There are numerous applications to theoretical computer science and the theory itself is very pretty. For instance, for any recognizable language *L* there is always a unique smallest automaton that recognizes *L*. The class of recognizable languages is itself capable of a number of elegant characterizations, some of which lead to the class arising naturally in unexpected places.

For readers wishing to experiment, you might like to try to draw an automaton that recognizes the following languages: (1) words that contain *ba* as a factor; (2) words that contain both the letters *a* and *b*; (3) words that end in the letter *a*. You may like to dream up your own, but you need to be wary—many simply described languages are not recognizable. For example, the language consisting of all *palindromes* (words that are themselves when spelt backwards, such as radar and minim) is not the language of any automaton: if A recognizes all palindromes, then it can be shown that it must necessarily also recognize some other words which are not palindromic.

I shall finish with a demonstration of such an unrecognizable language that allows us to make use of the Pigeon Hole Principle introduced in Question 7 of Chapter 6. The example is the language L of all words of the form $a^n b^n$, that is all words ab, $aabb$, $aaabbb$, ... (This amounts to saying that automata cannot count, or at least that they are limited as to how many things they can place in pairs.)

Suppose that A were an automaton that recognized all the words of the above language L. This is entirely possible, but I shall show that A will also be forced to recognize some words not of this type, so that the language of A is not L but some larger set of words.

For each number n, the word a^n takes A from its initial state i to some state we shall call s_n. Since $a^n b^n$ is accepted by A, the word b^n takes A from the state s_n to some accepting state, c. Now, since A has a limited number of states but there are infinitely many numbers n, it follows that there must be two different numbers, m and n say, such that the states s_m and s_n are the same even though the numbers m and n are not.

In view of this, consider the word $a^m b^n$ which is *not* in L because $m \neq n$. This word is, however, accepted by A as a^m takes A from i to $s_m = s_n$, and then b^n takes A from s_n to the same accepting state, c, as before.

Index

OXFORD

MORE OXFORD PAPERBACKS

This book is just one of nearly 1000 Oxford Paper-
backs currently in print. If you would like details of
other Oxford Paperbacks, including titles in the
World's Classics, Oxford Reference, Oxford
Books, OPUS, Past Masters, Oxford Authors, and
Oxford Shakespeare series, please write to:

UK and Europe: Oxford Paperbacks Publicity Man-
ager, Arts and Reference Publicity Department,
Oxford University Press, Walton Street, Oxford
OX2 6DP.

Customers in UK and Europe will find Oxford
Paperbacks available in all good bookshops. But in
case of difficulty please send orders to the Cash-
with-Order Department, Oxford University Press
Distribution Services, Saxon Way West, Corby,
Northants NN18 9ES. Tel: 01536 741519; Fax:
01536 746337. Please send a cheque for the total cost
of the books, plus £1.75 postage and packing for
orders under £20; £2.75 for orders over £20. Customers outside the UK should add 10% of the cost
of the books for postage and packing.

USA: Oxford Paperbacks Marketing Manager,
Oxford University Press, Inc., 200 Madison Av-
enue, New York, N.Y. 10016.

Canada: Trade Department, Oxford University
Press, 70 Wynford Drive, Don Mills, Ontario M3C
1J9.

Australia: Trade Marketing Manager, Oxford University Press, G.P.O. Box 2784Y, Melbourne 3001,
Victoria.

South Africa: Oxford University Press, P.O. Box
1141, Cape Town 8000.

Oxford
Paperback
Reference

OXFORD PAPERBACK REFERENCE

From *Art and Artists* to *Zoology*, the Oxford Paperback Reference series offers the very best subject reference books at the most affordable prices.

Authoritative, accessible, and up to date, the series features dictionaries in key student areas, as well as a range of fascinating books for a general readership. Included are such well-established titles as Fowler's *Modern English Usage*, Margaret Drabble's *Concise Companion to English Literature*, and the bestselling science and medical dictionaries.

The series has now been relaunched in handsome new covers. Highlights include new editions of some of the most popular titles, as well as brand new paperback reference books on *Politics*, *Philosophy*, and *Twentieth-Century Poetry*.

With new titles being constantly added, and existing titles regularly updated, Oxford Paperback Reference is unrivalled in its breadth of coverage and expansive publishing programme. New dictionaries of *Film*, *Economics*, *Linguistics*, *Architecture*, *Archaeology*, *Astronomy*, and *The Bible* are just a few of those coming in the future.

Oxford
Paperback
Reference

THE CONCISE OXFORD DICTIONARY
OF MATHEMATICS

New Edition

Edited by Christopher Clapham

Authoritative and reliable, this is the ideal reference guide for students of mathematics at school or in the first year at university. Nearly 1,000 entries have been added for this new edition and the dictionary provides clear definitions, with helpful examples, of a wide range of mathematical terms and concepts.

* **Covers both pure and applied mathematics as well as statistics.**

* **Entries on the great mathematicians**

* **Coverage of mathematics of more general interest, including fractals, game theory, and chaos**

'the depth of information provided is admirable'
New Scientist

'the style encourages browsing and a desire to find out more about the topics discussed'
Mathematica

POPULAR SCIENCE FROM OXFORD PAPERBACKS

THE SELFISH GENE

Second Edition

Richard Dawkins

Our genes made us. We animals exist for their preservation and are nothing more than their throwaway survival machines. The world of the selfish gene is one of savage competition, ruthless exploitation, and deceit. But what of the acts of apparent altruism found in nature—the bees who commit suicide when they sting to protect the hive, or the birds who risk their lives to warn the flock of an approaching hawk? Do they contravene the fundamental law of gene selfishness? By no means: Dawkins shows that the selfish gene is also the subtle gene. And he holds out the hope that our species—alone on earth—has the power to rebel against the designs of the selfish gene. This book is a call to arms. It is both manual and manifesto, and it grips like a thriller.

The Selfish Gene, Richard Dawkins's brilliant first book and still his most famous, is an international bestseller in thirteen languages. For this greatly expanded edition, endnotes have been added, giving fascinating reflections on the original text, and there are two major new chapters.

'learned, witty, and very well written . . . exhilaratingly good.' Sir Peter Medawar, *Spectator*

'Who should read this book? Everyone interested in the universe and their place in it.' Jeffrey R. Baylis, *Animal Behaviour*

'the sort of popular science writing that makes the reader feel like a genius' *New York Times*

OXFORD LIVES

'SUBTLE IS THE LORD'

The Science and the Life of Albert Einstein

Abraham Pais

Abraham Pais, an award-winning physicist who knew Einstein personally during the last nine years of his life, presents a guide to the life and the thought of the most famous scientist of our century. Using previously unpublished papers and personal recollections from their years of acquaintance, the narrative illuminates the man through his work with both liveliness and precision, making this *the* authoritative scientific biography of Einstein.

'The definitive life of Einstein.' Brian Pippard, *Times Literary Supplement*

'By far the most important study of both the man and the scientist.' Paul Davies, *New Scientist*

'An outstanding biography of Albert Einstein that one finds oneself reading with sheer pleasure.' *Physics Today*

PHILOSOPHY IN OXFORD PAPERBACKS
THE GREAT PHILOSOPHERS
Bryan Magee

Beginning with the death of Socrates in 399, and following the story through the centuries to recent figures such as Bertrand Russell and Wittgenstein, Bryan Magee and fifteen contemporary writers and philosophers provide an accessible and exciting introduction to Western philosophy and its greatest thinkers.

Bryan Magee in conversation with:

A. J. Ayer	John Passmore
Michael Ayers	Anthony Quinton
Miles Burnyeat	John Searle
Frederick Copleston	Peter Singer
Hubert Dreyfus	J. P. Stern
Anthony Kenny	Geoffrey Warnock
Sidney Morgenbesser	Bernard Williams
Martha Nussbaum	

'Magee is to be congratulated . . . anyone who sees the programmes or reads the book will be left in no danger of believing philosophical thinking is unpractical and uninteresting.' Ronald Hayman, *Times Educational Supplement*

'one of the liveliest, fast-paced introductions to philosophy, ancient and modern that one could wish for' *Universe*